COWS

Credits:
COWS
A Novel By
Matthew Stokoe
ISBN 1 84068 005 9
First published 1998
by Creation Books
This first paperback edition published 1999
by Creation Books
Copyright © Matthew Stokoe 1997
All world rights reserved
Design:
Bradley Davis, PCP International
A Bondagebest Production

Author's dedication:
For Roseanna, In Memoriam

COWS

A NOVEL BY

Matthew Stokoe

Chapter One

In bed.

Steven could feel the toxins tumbling slowly through his bloodstream, jagged black particles that rolled in a slow-motion undersea current, gouging soft tissue with their passing. If he closed his eyes in the dark room he could see a science-book photo of his blood map. Blood, not sticky red liquid, but billions of corpuscles all back-lit in a fireside glow, jostling for position in a race to the heart that would love them and pump them down to the lungs for that good, good oxygen. The heart wanted them to live and it cheered on its team with the unshakeable, endlessly enduring love of TV parents.

But riding the backs of his corpuscles, leaping on to them from his stomach wall and through the slick grey coils of his intestines, not giving a shit what his heart wanted, the hard black grit of Mama's catabolised meals jammed itself into his flesh and fat and gristle.

On his back in the greasy ruck of bedclothes, he could feel the thousand systems of his body clogging with this filth.

He turned on his side and looked out at the 3AM city through a single uncurtained window. It didn't work. In the cold gutted room, on the narrow bed wedged into the grudging protection of a corner, he could still feel his body age.

He raged at his powerlessness. She forced her mealshit into him day after day and he couldn't stop her. He wanted to. He wanted to tie her legs apart and take a hammer to her cunt then walk out on to the street and never come back. But he couldn't.

In the long nights before sleep the TV had no pity. It showed him how the world was. It showed him how much the people out there had. He'd been outside to see for himself, of course, out into the city and walked around. But it was too frightening to stay out for long. He wasn't like the people on the streets. They lived so perfectly. They knew exactly what to do to be happy and they did it without even having to think. And the TV beamed their lives into his head as dreams.

Across the bare floorboards, in a patch of sick orange light, Dog lay sleeping, its paralysed back legs out stiff like the handles of a wheelbarrow. Steven closed his eyes. At the edges of roads all across the world sodium vapour lamps sizzled away at the night, and in the flat upstairs the new girl moved around and spoke to herself.

Matthew Stokoe

Chapter Two

In the mornings, if the water ran hot, Steven could stand for hours in the raw concrete of the shower stall. Like sleep it was an escape. The flow of the water soothed him, threw a cover over his emptiness. It was like the few times he had ridden a bus – without doing anything you were doing something, you were moving, and the movement absorbed you. All the headchatter went quiet and you could imagine you had all the things you saw on TV, like love and a ranch in the forest with a horse and a brand new Jeep and a child and a wife who loved you and would stroke your cheek when you got home so tenderly that you knew she lived only for you and when you walked through the forest or the city a path opened up and you always knew which direction to take and nothing ever jumped out and stopped you or cut you off from life because you were right in there with it, you were part of it all and you didn't miss out on a thing. And when you looked at the TV it was a mirror.

But when Steven stepped out to dry himself with a rag, when his feet hit the scummy stone floor in front of the toilet, nothing had ever changed.

Gargantua. The Hagbeast. The unloving mother bitch cunt. Stood hulked over a two burner stove, stirring a pan of rancid pork. The kitchen stank of gas and oil and the caked, dead fish decay that came out between her legs.

Steven sat at the small unbalanced table and watched Dog drag itself across the sticky lino to the shit tray. Its useless hind feet swished sideways with each lurching foreleg step, like the tail of some broken fish. He'd had it from a pup – nine years – and had been there, standing impotent and frozen, when the Hagbeast crippled it with a brick. For no reason at all.

That day in his teens was confirmation of what he had suspected since birth – that he was incapable of manipulating life as other people did. Unlike them he could have no effect on the web of events that surrounded him, he could bring about no change. Dog had looked up at him not savage or pain-snarly but confused, like how could Steven allow this to happen?

In those days Dog was young and had not learned how powerless Steven was before the Beast.

Out in the hall now, the animal shitted a dark turd into a bed of ripped newspaper. Good boy. Snapped in half and still killing itself to please.

The Hagbeast brought breakfast over.

"Here, lovey. Mama's best boy eat that all up."

She sat opposite him and slopped chunks of undercooked meat on to his plate. The oil that soaked it was flecked with something that looked like phlegm.

"Eat up, eat up. Got to eat Mama's food that she makes just for you, haven't we?"

Steven looked at the sagging face, at the cross-hatched pouches of fat and the clogged skin, at the ancient blackheads that had grown with the years, outward like the rings of a tree. The grey hair on her jowls lay flat under the crusted remains of a thousand meals and she had snot on her upper lip. He summoned his courage.

"I can't eat this."

He prodded the food with his fork, dropping his eyes, wanting to challenge her but unable to bear the terror of her gaze. The Hagbeast sighed and her voice got hard.

"Everyday. Everyday the same fucking thing. I made this food with love, Steven, and I want you to eat it."

She made a fist around her spoon and shovelled in the slop. Her movements were ponderous and inexorable, as though some highly torqued mechanism revolved beneath the lax obesity of her frame. Fat swung in pleats from her upper arms and she breathed heavily through her nose as she chewed.

"It's shit. It isn't even cooked properly."

The Beast spat out a mouthful of food and started to shriek.

"Shit! Shit! You ungrateful fuck. People out there would die to eat this."

Steven held tight to the leg of the table and pushed his words out like small boats into the storm of her screaming.

"Food like this kills people."

"Eat the fucking food!"

Her words rang on the filthy tiled walls. In this tight space outside the world their fury silenced the city. She heaved herself upright and stood waiting, daring him to refuse, grunting low in her throat and pressing her teeth together.

Steven didn't have the strength to resist further. His fear of the monstrosity before him withered to dust the small store of opposition with which he'd hoped to transform the morning. He speared a piece of the meat with his fork. His stomach rolled, but like every other mealtime he filled his mouth and chewed and swallowed. And kept on doing it until his plate was empty.

Chapter Three

On the bus to the meat grinding plant the cereal and fruit fed faces of the other passengers made him feel haggard and polluted. He wanted to reach out and touch them, to reassure himself that he belonged to an essentially similar world. But he knew he did not, and that if he tried they would telescope backwards like some effect on TV.

He watched them instead. They were so much more real than himself, the air around them was bright with the definition of their existence. He felt himself blurring with the sunlight and the motion of the bus, as if his outline were sand or fine powder.

There were couples, too, together on the slashed seats, and they were the most densely coloured of all. Their belonging, their completeness, pushed them out from the background of safety glass and pressed steel, up so close to Steven that he could feel the flow of love between them. These were the ones whose lives got shown on TV. They knew the secrets of the game and they played and never considered losing.

They were gods from some golden other world.

They had arms, legs, their faces moulded to their emotions as his did, they even aged. But they were beyond him. The air they breathed was not his air and the light that fell on them came from a warmer source than his sun. He longed to imitate them, to share in the mass normality that rolled in cathode waves across the dead nights of his loneliness.

The bus was almost empty by the time Steven stepped from it into the deathstink at the edge of the city.

Chapter Four

The meat plant squatted low in a gritted wasteland of industrial units, hunkered down and curled like a bellyshot animal. Smoke and steam coiled out of pipes in its sides and pools of water in the fractured concrete apron collected a scum of oil and condensing cow fear that reflected the jaundiced sky back at itself.

Articulated trucks arrived endlessly. They pulled up at the stock pens spewing shit and black exhaust and emptied themselves of cows that farted and mooed and jerked around trying to remember if Mom ever said anything about a place like this. But there wasn't much time for remembering, the pens were in constant flux, draining at four animals a minute into the plant, through a hole in the wall.

In the front office they gave him a white coat and a cap and cream rubber boots that looked like tanned gut. It was his first day and he had to be properly dressed.

There was a lot of noise and people said things to him, but he didn't speak unless he had to. He was adrift in their world, unsure of his significance, and to open himself to a point where conversation could take place would only have revealed how unlike them he was.

Cripps lead him through office corridors where the air was guilty with knowledge of the killing out back, and as they moved deeper into the plant, further from the administration section, things changed – the temperature fell, there was less light, the staff thinned out and those who remained looked harried and dark-eyed.

"Pussies."

Cripps spat on the carpet-tiled floor

"The whole fucking lot. They sign papers and shuffle them around and a ton of meat dies every minute. But not one of those syphilitic cunts has ever had his dick in a cow. They don't know what it means to slaughter through an eight hour shift, to kill and keep killing until the death of an animal sings to you of things beyond yourself."

Steven followed the foreman, not really listening, too busy sucking in details of the scene around him to match later against the TV – jewels of actual experience to be taken home and gloated over.

They came to a corrugated iron wall that stretched thirty feet up to the roof and off to the lost edges of the building. Cripps held open a door in it and the white light that streamed through blinded Steven and made the men on the other side look like angels in some kind of movie about heaven.

"This is where things are real."

Cripps shoved him into the light.

Steven stood blinking beside a processing line that curved round three sides of an immense hall. Carcasses, hung from hooks in an overhead conveyor system, swung upside down into the hall through a square of plastic strips at one end. Wet shit slopped down their flanks and blood dripped from their noses into the polished steel gutters that tracked the line. Process hands in blood-stained white coats tended various stations, washing down the heavy dead cows, slitting them open with small circular saws, scraping out guts, skinning, slicing, hacking, boning, dismembering, rendering the once solidly knit animals into chunks of unsupported flesh. The keening of electric knives as they parted skin and meat cut holes in the coarser whine of bone saws and the repeated crumping of a pneumatic skull press. Cripps, his hand on Steven's shoulder, had to shout.

"That's the best part over there, boy. The slaughter room."

He pointed to the curtain of plastic strips that marked the start of the dead cows' journey.

"But we're starting you off on the grinder."

The process hands paid no attention to Steven as he followed Cripps, but he watched them closely, imagining the lives they must go home to after work, all their beautiful wives and children.

"This is the end of it all."

They stopped beside the stainless-steel chute of a machine. From a conveyor-fed work surface Cripps took a slab of beef the size of a small child and heaved it in. Bits of meat sprayed back, but the bulk of it, ground to a pulp of blood and tissue, gouted out the other end into a wheeled hopper. He scooped up a handful and rubbed it through his fingers, standing with his groin pressed to Steven's hip.

"Look at it, boy. We haven't just killed it, we've obliterated it."

He smelt his fingers.

"Think whatever made it move is happy now in the fields of the hereafter? You believe in that kind of thing? Forget it. Meat doesn't have the brains. It just works till it dies or until someone cuts it up."

Cripps looked dreamily across the hall at the juddering procession of increasingly disintegrated cattle.

"Just chuck the shit in as fast as you can."

He squeezed the back of Steven's neck, then strode off towards the slaughter room. Steven watched him go.

The meat juice stung his hands after a while, but there wasn't much else to bother Steven. His shoulders ached slightly with the effort of humping the meat but the motion was rhythmic and simple and he lost himself in its mindlessness. He dreamed he was working to provide for a beautiful wife and a baby son. They waited for him at home with two cars in a quiet neighbourhood where all the houses had large lawns. They depended on him and the wife would be wondering how he was doing at work and talking to the son about how much she loved his father, glowing inside with the knowledge that she would never change, that no other man could ever mean anything to her and that she would always live only for Steven. And she had a good body too, really nice firm tits but not too big and skin like all women had under the warm painting of TV lights – lightly tanned and smooth as silk.

At 1PM he got a break and wandered round the line. After the cows were skinned and gutted they were decapitated and the heads routed off to the skull press. Gummy operated this machine like a personal weapon, as though the steel pile that slammed down on the dribbling heads, splitting cranial vault and exposing brain in a thick spray of colourless liquid, functioned for his satisfaction alone. He groaned and pressed his knees together each time he triggered it.

Steven looked at the man's crotch expecting an erection.

"You starin' at mah mouth?"

The words flapped over Gummy's chin like drool. Steven looked at his mouth – no teeth, lips torn away, the left side of his face an open purple scar that showed gums and leaked spit.

"You starin' at mah mouth, ya little bastard? Every new cunt does, and I can see you ain't no different. Bet you're just dying to hear how it happened, ain't ya?"

Gummy kicked another head into place and hit it dead centre. Some of the stuff that came out of it stuck to Steven's coat. He thought of walking away, but there wasn't anywhere to go.

"You wanna watch these cows, I'm tellin' ya, ya little bastard. I had mah hands holding right on the soft part behind the ears, just where they like it most... God, the skin there is like silk... And I had my lips right up against that mad bovine's. I could feel the whiskers and the velvet under those whiskers all dark and smelling of hay. So just like always I opened mah own lips and I tasted that cow. I felt its tongue pushing into mah mouth and I pushed back against it – they're real rough on top, but underneath you never felt anything so smooth. Anyhow, I was licking at that scummy stuff they get on their teeth when the mad fucker pulled back and clamped down on mah mouth and started shaking its head. They had to pry me loose with a crow bar, killed that bastard cow to do it too. By that time mah teeth was gone and mah lips was ripped up so bad they never got found. Like it, ya little bastard? Bet you think it's a real good story, dontcha, ya little fuck? Well you better remember it. Cows taste like heaven but ya can't trust 'em a single fuckin' inch."

Gummy started work on another head and Steven went back to the grinder.

Chapter Five

Outside his building, in the dark. Sodium vapour drained colour from the blackened walls of the ruin and Steven found it hard to focus on the crumbling planes. Everything was too bright or too black and the wash of streetlight ochre fritzed his eyes. The place had a shunned look, as though whatever gave personality to a structure had flown in disgust from this one long ago. And, left abandoned in a field of belching farting puking factories, the rotting four storey Victorian had grown shuttered and autistic.

Steven paused on the steps and watched the evening trucks blow litter along the empty sidewalk, and wondered where they were going.

In the gloom of the third floor landing, as he stood at the door of the flat steeling himself against the Hagbeast, he felt the unfriendly air move. Darkness swirled on the stairs then parted and she was walking towards him in slow motion strides. Lucy – black t-shirt, black leggings, dark hair streaming in some self-generated slipstream. Half Indian, half Jewess. She ran her eyes over his face like a blind woman using her fingers, concerned not to communicate immediately, but to search for possible hostility. Steven stood passive while she picked a piece of ground meat from his hair. She held it on her palm and stared at it.

"I do that at work. It's my new job."

She lifted her gaze and looked intently at him.

"When you cut them open do you get to see inside? Do you reach inside and look around?"

Steven shifted his feet.

"For what?"

"They live, don't they? They suffer. Like us. Haven't you seen the poison inside them? Hard and black and stuck in the intestines? Or under the liver or somewhere else?"

"I just saw guts. Toxins are stored in muscle, anyway. Not in lumps between organs."

"I'm not talking about toxins. You think I'm talking about sugar and caffeine and all that shit? Fuck, just being alive does it. That and what your parents do to you before you get strong enough to stop them. And even when you can stop them it's too late. The seed's there and it grows and grows until it jams all the systems in your body and your mind fucks up. Didn't you see anything like that?"

She sounded desperate.

"I didn't really look, maybe there was something."

Her intensity made Steven uneasy, but she was a woman, a possible source of love, and he didn't want to disappoint her.

"The only thing I saw up close was the heart. They weigh six pounds, you know. It was still beating when I held it, like it was trying to suck something in. But it stopped in the end."

Lucy looked suddenly tired.

"Our hearts are only two pounds, not much room for love."

He watched her drift back up the stairs to the fourth floor, watched her brown fingers trail along the bannister and imagined them on his cheek.

The flat was cold. The Beast in her gluttonously accumulated blubber did not feel it. Steven went straight to the bathroom, he needed a shit. He could feel it heavy in his guts, squeezing closed the large veins in his groin, making his legs ache. Minute fecal particles would be sifting through the wall of his colon, heading straight for the cells of his face and brain, ageing, ageing, ageing... stealing the future.

He locked the door, squatted, flipped his dick inside the toilet seat, leant forward and strained. After a moment of pissing and resistance his sphincter relaxed and a foot and a half of grey shit shot out of his arse and left a thick smear on the dry porcelain above the water. Steven twisted to look at it. Although it was large, one end was broken raggedly across and he knew his dump hadn't been complete. It never was, his body never managed to get rid of all its poison in one go. He wondered if this was what Lucy meant.

He used a lot of newspaper wiping himself.

Chapter Six

In his bedroom. Dog scraped across floorboards to snuffle hello. Steven petted him sadly. A dog was such a symbol, it meant so much on TV and in the world outside. It was connected with rambles through sunlit meadows, carefree and laughing, arm in arm with a loosely dressed woman, throwing a ball for a delightfully squealing child to chase. But Dog knew little of sunlight. This wreck of an animal had lived out its life without ever once having escaped the shadows of the flat.

Something heavy lumbered along the passage outside – the Hagbeast emerging from the rear of the flat, snorting her way to the kitchen like a pig rooting its way through a dung heap. He could picture her exactly – head lowered and forward, nostrils wide, spit stranding from her chin to the filthy floral print material of her breast. And the rear view – a blot of wet menstrual blood sticking the dress to her rolling arse and the backs of her thighs, hunched shoulders, bare mottled calves, swollen like the rest of her. Even through the closed door and the peeling walls he could feel the effluvium of her hate. He wondered if she could feel his, it was just as strong.

It had never been different. From the second she had squirted him from her cunt they had loathed each other. In the littered kitchen, on the table they ate from still, she had pulled him out of the mess between her legs and cursed him. And he, sensing a lifetime of worse to come, had pissed in her eyes.

Steven did not step outside the flat until he was five. By then, though the swelling of his heart at the unimagined largeness of the world told him to run as far and as fast as his little legs would carry him, he was enough aware to understand that he could not survive alone. The Hagbeast was, for the time being, necessary to his existence. But from that moment of glimpsed possibility his child brain started to count the months to maturity and escape. For each year that passed afterwards, there was a year just coming that would carry him into self-sufficiency and freedom.

But it didn't happen like that. By the time thirteen and fourteen and fifteen came (and all the others), even though his hate for the Beast and for his squalid, closeted life had in no way diminished, he found he had somehow left things too late. His five year old fearlessness had atrophied to a point where it was impossible for him to contemplate extended periods beyond the walls of the flat. Through the years of his growing the Hagbeast had so leeched him of those identifying marks by which the world might possibly have recognised him that escape by simply leaving this place had become a laughable notion.

Steven stayed in his room as long as he could, sitting on the bed, idly stroking Dog while TV images fluttered into the room like whores' promises. But eventually it came, as he knew it would – the double-tracked horror movie screech that drew tight the reigns of her proprietorship.

"Steven!"

His skin crawled.

"Steven, dinner's ready."

If he waited any longer she would come for him, so he stepped into the hall and trudged to the kitchen. Dog grunted along behind him.

He knew immediately that things had changed, that there had been a shift in attitude. Small things – the way she stood and looked at him, a subtle rearrangement of her fat, even the shape of the blood on the back of her dress – a thousand hints that marked the beginning of a new phase of misery. Steven moved warily to the table and sat down, keeping his eyes on her.

"You didn't want to keep Mama waiting, did you?"

"I was tired."

"Of course you were. There."

She put something in front of him. Steven looked at it in disbelief – part of a sheep's stomach, steaming in folds that hung over the edge of the plate. It had not been cleaned and undigested vegetable matter speckled the frilly corrugations of its inner surface. He touched it with a finger.

The Beast, already chewing, noticed the movement.

"I know you like this. I made it special so you could have a nice dinner after your first day at work. Go on, start."

Steven didn't move and the Hagbeast grinned at him.

"Mmmm. It melts like butter. Hurry up, don't let it go cold."

"No."

"Oh, yummy, yummy, yummy. I've cooked you up a treat here. Eat, eat."

The sing-song in her voice worried him, there was a deadliness behind it. Things were escalating.

"I said no. I'm not eating it."

The Beast put her spoon down slowly.

"And just exactly what is wrong with it, Mister Cocksucker?"

"People don't eat things like this. People don't cut something out of an animal and put it straight on a plate. It's not clean."

The Hagbeast choked on laughter and blew snot and tripe across the table.

"Oh, people. Peeeple. Look at the cunt, such an expert now. Ooo, a whole day out there. You must know everything."

Steven squeezed his fork until it hurt his hand.

"You're a fucking idiot, Steven. You think going out there for a day makes you like them? You think you got strong today? Show me how strong you are, cunt. Walk out of here and find somewhere to live... You moron. Without me, without this home I've given you, how long would you last?"

He felt his guts liquify. He wanted to scream at her that he could be like them, that one day he would have love and a wife and everything else. But he knew the bitch was right, he couldn't leave. He had plans. He needed safety in which to copy the lives he saw on TV. The dreams he had of rebuilding himself would be impossible to fulfil without it.

"You get too cocky, you little cunt, and I'll put you out myself. How would you like that? All those people around you for ever and ever and no place to get away from them. Wouldn't that be fun?"

"No."

It was hard to breathe, his chest felt too tight.

"What was that? Mama didn't hear you."

"It wouldn't be fun."

"No it wouldn't, would it? So eat your fucking food."

Matthew Stokoe

Steven cut into the organ on his plate and put some of it in his mouth. It took forever to chew. The rubbery flesh slipped around his teeth and made him heave.

"Yes, yes, that's a good boy. That's Mama's good boy, eating it all up."

But Steven wasn't listening. In the middle of a vomit-ocean of misery he was busy deciding on the meaning of her words.

The flat was hers to take away as she wished. This had always been so, but she had never used its removal as a threat before. So why now? Had the bravery of his first day at work made it plain to her that he had hopes for the future? If this was so he would have to be careful, the bitch would kill him for sure if she thought he might escape the hell she had spent so long building around him. Perhaps the upscaling in food disgustingness was her first move.

He chewed on and forced the sheep's belly into his own. The Hagbeast squirmed to unstick her bleeding arse from the chair.

Steven watched TV a long time that night, searching the scattered pixels for some way to armourplate himself against the Beast. On screen the templates for life were easy to find, but the methods of their construction, as always, remained hidden.

He'd puked his dinner in the corner of the bedroom and Dog had eaten it. Having been inside its master's body the offal and bile were sacrament to the animal, and their consumption scorched its brain with dreams of becoming man. The air was still sour and the soft membrane behind Steven's nose burned.

Upstairs Lucy moved, creaking floorboards as she walked from somewhere to somewhere else. Steven imagined what he would see if she was naked and the ceiling was glass.

He probed his guts with stiff fingers to see if he could feel anything hard and poisonous that might be the cause of his abnormality. But all he found were vague outlines of functional meat that made him flash back to swinging cows with veined sacks of organs spilling across their chests.

He stared at the black ceiling, unable to sleep. He heard the rusty sponging of Lucy's bedsprings and saw himself waiting in a double bed in a wide, pine panelled bedroom filled with birdsong and her just getting in. He felt the compression of the mattress under her weight, the slide of her dark skin against his, the moulding to his body... and the relaxation of love.

Then Dog yelped out of a dream and it was quiet upstairs and Steven's body sagged against its own emptiness.

And the night dragged on.

Matthew Stokoe

Chapter Seven

Breakfast was bad, it followed dinner's pattern. The Hagbeast stood over him while he tried to eat, clamping his head in the crook of her arm and forcing the food into his mouth with her fingers when he wasn't fast enough to please her.

"Did you dream about me last night, Steven?"

She pressed her mouth to his ear and her breath stank. He could hear the spit collecting in the back of her throat.

"I found come on your bed this morning. Dog was going to lick it up, but he didn't when I came in. It was thin. You need more of Mama's cooking, that stuff spread on my hand like milk. Mama wants you to be strong, doesn't she. She wants your come all thick and gooey. Eat up, there's a good lad, oh that's a good boy. Swallow it down, that's right."

Steven jerked his head free and wiped his face.

"You fucking mad bitch. I didn't come and I didn't dream about you."

"If you say so, but I think Mama knows what she's

talking about."

The Hagbeast sat opposite him and started on her own food.

"See? I don't know why you make such a fuss. I eat just the same as you. Whatever my boy eats, I eat. That way we'll both be strong, won't we?"

She forced a mouthful of food through the gaps in her teeth, out past her lips in a glistening wad, then sucked it back in and swallowed it.

"You don't want me strong. You're killing me with this shit."

"Oh Steven, please."

"Look at my skin, it's grey."

"All boys have trouble with their skin."

"I'm twenty fucking five."

The Beast ground her teeth.

"I know how old you are. Believe me, I've counted the years. I've counted them and I've seen them fill up with your disgusting habits. Do you know how bad you stink when you take a shit?"

Steven's rage choked him. He wanted to kill her, he wanted his body to explode and destroy the suffocating little kitchen. But it wouldn't. It sat there paralysed by a carefully inbred fear of action against this fat hulk and did nothing. Only his mouth seemed to be working.

"How can you smell anything over that mess between your legs?"

The Hagbeast knocked over the table and stood up, jowls quivering, fists pushing into the fat of her hips.

"You fuck. You shitting fuck. How dare you talk about my blood. My blood is the marker for the wound of your birth. It never healed, Steven, and I let it run so I never forget the horror of that day. You pissed on me, you jackal. I should have killed you then...."

And on and on until Steven broke from the table and ran out of the flat.

Chapter Eight

The men on the line stood around smoking and drinking coffee out of plastic cups, breath steaming in the refrigerated air of the processing hall. They joked and talked about women, squeezed their balls, draped arms across friends' shoulders. Even killing time, they were more alive than Steven ever expected to be.

He sat on his stool at the grinding station waiting for the shift to begin. There was no-one near him so he stared at the ventilation grilles set into the walls down near the floor and wondered what it was like behind them. Over at the skull press Gummy dripped oil on to moving parts.

Two minutes before horn time, six men in spotless white overalls marched purposefully across the hall and through the hanging plastic strips of the slaughter room entrance. They moved with a vigorous precision and they shared nothing with the other process hands. They were lords, apart and above, the forces of the world meant nothing to them. Steven tracked them until he felt a calloused hand on the back of his neck.

"Slaughter party."

He twisted to find Cripps behind him, staring wistfully after the men.

"God, what beauty...."

He looked hard at Steven.

"They are the genesis. They create what the rest of you only work with. They know themselves, boy. They looked inside and weren't afraid to drag out what they found. What would you find inside yourself, I wonder?"

The hand on his neck moved to rub his shoulders.

"We all have it, that dark core. It makes us men. And if we examine it, if we can bear to hold it up to ourselves and acknowledge it as our own, then it makes us more than men. The slaughter room is where we become complete, boy."

Cripps gave him a final squeeze and walked off, straight backed and clear eyed. The horn sounded. Cups and butts hit the floor. Process hands moved to their stations and dead cows started to swing through the plastic strips.

Steven ground meat all morning, absorbed in the speed and power of the grinder. The machine blasted out its meaty pulp with such force that it stuck to the sides of the hopper and he tried to slow it down by overloading it with the heaviest, toughest chunks of beef. It didn't work, all differences of size and texture became uniform under the spinning discs and spiked rollers.

By midday the fascination had worn thin and Steven was on auto, twisting from the waist to grab the meat, jerking it up, letting momentum carry it at the end of his arms into the mouth of the grinder, then twisting at the waist again, starting a reverse arc for more meat... and humping it up and dumping it in.

His gaze wandered over the mechanised carnage of the line, lingering on hands slicing with electric knives, on the bodies falling swiftly to bits. The fact that organs were piled into carts and limbs sheered from carcasses did not revolt him so much as make him aware of his own mortality. How easily he could suffer, at any time of day, some accident that would burst him or crush him or mangle him before he ever got a chance at happiness. And if not an accident, then the Hagbeast's poisoning. He was sure that last night she had passed some pinnacle of restraint and was now coasting towards a time when he would lie cold and stiff across the kitchen table, the remains of a last incrementally fatal meal squashed against his chest.

Something flickered at the edge of vision, a shadow in wire mesh. He snapped round to stare at the ventilator behind him, but beyond the grille of steel wire there was only darkness. Before he could slide off his stool to take a closer look the lunch horn sounded and the walkway between him and the wall was suddenly full of men racing to the locker room for sandwiches wrapped in tin foil by doting wives. So Steven forgot about it and waited for them to pass, and when the process hall was empty he walked round to the gutting trough, an early station not far from the slaughter room.

The line was still. A cow swung by its heels from the overhead track, going nowhere. Its tongue lolled softly from side to side and strands of saliva traced patterns in the blood of the run-off gutter. Steven reached out to stop it swinging and left his hands on the wide sides of the animal for a moment, feeling ghosts of life in the cooling skin.

The belly was slit and the sack of guts, not yet removed, had fallen out to hang heavy and mottled against the ribcage. He breathed deeply through his nose, searching for some scent of the grassy, wildflower-strewn field in which this animal must have grazed. But the dusty smell of dung and hide and the fermented, rotted-down stink of the exposed organs blanketed everything and Steven had to

Matthew Stokoe

close his eyes and force himself to imagine the beauty of where the cow had lived.

He found a curved gutting knife and used it on the animal. He had to reach inside the abdominal cavity and the unexpected warmth he found there pulsed a brief wave of sympathy through him. But it passed quickly.

The guts slid over the cow's stretched neck and landed at his feet with a sound like someone being sick on a tiled floor. He stood there trying to recognise the different parts. The heart and lungs hadn't come out of course, they were still fixed within the chest, but the spleen and the kidneys were there. And the enormous liver and, most easily identifiable of all, the tangle of grey-blue intestines weaving slickly about themselves, shiny in the harsh light. Among these landmark organs were smaller, irregularly shaped bits of viscera he couldn't name.

There was very little blood. The cows died quickly and trapped most of it in their tissues, a last snatching back of themselves from all the touching hands of man. Dark bile leaked from the ruptured stomachs, though, and a shallow fringe of clear internal mucus collected around the edge of the pile. Steven crouched and examined things – the hollows, the tight bunching of hard yellow fat in the dip of the kidneys, the smooth brown slope of the liver, the pockets of viscous pink glit....

The mess was incongruous, there on the hard floor, but within itself it was consistent, all of it grown to a single plan. There were no crystalline black accretions jammed into organically curving tissue. It looked like Lucy was wrong, at least as far as cows were concerned. But he had to be sure.

The touch of the organs when he stuck his hands into them was unpleasant. Instead of the softness he expected, he found them hard with vaguely abrasive surfaces. He rummaged quickly, running his fingers along folds and crevices, poking through valves and into sphincters, probing the insides of those that would admit him.

Slivers of meat collected under his fingernails and everything made wet sucking noises. He was thorough but he found nothing to comfort Lucy, nothing she could use as proof.

"Looking for God, boy?"

Steven jumped and the leathery bag of one of the cow's ancillary stomachs slipped from around his hand. Cripps stepped forward, smirking, and stirred the entrails with the toe of his boot.

"Marvelling at His creation?"

"What?"

"It's food and shit, boy, that's all."

"I was looking for something...."

Steven's voice trailed off. He was frightened, unsure what Cripps' reaction would be to his sifting of guts.

Cripps laughed and put an arm around him.

"Then perhaps I can help you."

He steered Steven to the slaughter room entrance and paused for a moment outside, savouring some quality in the air. The plastic strip curtain blurred angles and lines and muted the cow noise beyond to a nervous grumble.

"Come on."

Cripps was gentle and they moved into the room.

Steven had expected a cathedral to death, but the raw concrete cavern seemed squalid and mean despite its yardage. At the back was a holding pen of dull steel, fed from the stockyard outside. The cows here, waiting for the return of the slaughter party, rocked sideways on their hooves, chasing cow lullabies across the dead-eyed plains of their pasts. But Ma's mud-soft lowing was too far back across those plains to give comfort and the cows were cold.

Matthew Stokoe

From the pen, two barred alleys ran to pneumatic grids the men called grabbers – rigid latticeworks of iron that closed against the sides of a cow and held it immobile. On either side of these there were low railed platforms for the slaughtermen. Over the entrance a winch connected with the overhead conveyor.

In this place of bovine departures the lighting was dim. Alcoves and juttings, thrown with no seeming purpose along the walls, wrapped small areas of darkness about themselves. At the top of a flight of steps a shelf of stone ran the length of the room, ten feet above the floor – a viewing gallery.

"Look around you, boy."

Cripps spread his arms.

"It's quiet now, but you can feel the power of the place. Think of the deaths it has seen, the fantasies that have been lived and released in here. God, that smell...."

Cripps walked along one of the alleys to the holding pen and stroked the forehead of a cow. He raised his voice and the animals shifted uneasily.

"These are your future, if you have the courage. They grow them in concrete boxes under ultra violet light, they feed them on pellets of their own dead. These are urban cows, boy, man-made without mystery, and they have a gift for us far more important than meat or leather. It isn't a gift they like to give, though. Not at all."

"What gift?"

"The experience of killing. Of blowing out their brains and taking away their most precious thing. It smashes the walls you put around yourself, the walls other people put around you to stop you doing what you want. Do you understand me? The things you would do if there was nothing to stop you. Killing is an act of self-realisation, it shows a man the truth of his power. And when you know this, boy, the pettiness they try to shackle us with falls away like shit."

Cripps threw his arms out like he was on a cross.

"Killing frees you to live as you should."

Out in the hall the horn blasted.

"Back to your station, boy. Back to where the cows are only meat. But remember what happens in here, remember the secrets that are to be had. And one day soon perhaps we shall see what a little killing can do for you."

At the grinder Steven humped meat and dreamed of quick access to the future. Cripps was significantly fucked in the head, no doubt about it, but could it happen like that? Was there something you could do that would make you different than you were? If it was that simple, how easy it would be to deal with the Beast.

His head swam a little and the mist of blood from the grinder began to irritate him. Lucy with her compacting of unhappiness into removable physical deposits, Cripps and his instant command of life through killing.... Such new ideas. Steven had not thought that there might be ways to force happiness into being. It had always seemed a matter of luck, something beyond his control that happened outside in the world. To all the other people.

He moved about, uncomfortable in the late afternoon slaughter. Someone was watching him, he could feel it. But he was apart from the other process hands and Cripps had not left the slaughter room since lunch. He looked over his shoulder. In the darkness behind the ventilation grille two softly gazing eyes blinked once then vanished. He jumped from his stool, but it was too late, the space behind the grate was empty. He pressed his head close and from somewhere along the duct's length heard a sound like lazily trotting hooves.

Matthew Stokoe

Chapter Nine

Dinner looked normal that night – junk out of a can. The Hagbeast ate silently but watched him closely. The first mouthful told Steven she had laced the meal with salt. He forced himself to eat without reaction.

"Is it nice where you work?"

"No."

"When I was a girl I worked on a goose farm. That was bloody work, too. They put them upside down in tin cones with holes in the bottom. There was nothing in the sheds but rows and rows of goose heads hanging out of cones. We had to run along with a knife cutting them off and the blood went everywhere. We were always soaked. They looked like cocks those heads did with their long necks, lying all bloody on the ground."

Steven's stomach jerked. Her words had no effect on him – he had heard all her stories before he was eight, how she used to stick the necks inside herself – but the salt was building with each mouthful and his guts were going to empty sometime soon. He forced more food down to spite her.

Out in the hall Dog dragged itself up for a shit. Steven flicked back to the Hagbeast.

"Don't bother, I've heard it."

"Oh, well, I'm sorry. I beg your fucking pardon. Mothers are supposed to talk to their children, Steven. Didn't you know? It's the only way to teach you things."

He laughed in her face.

"What did you ever teach me?"

The mother-act fell from the Beast like lizard skin and she leaned across the table, grabbing the edges, white knuckled.

"You ungrateful fuck. Everything you are has my mark on it."

His stomach heaved again as he half stood to meet her, but he wasn't ready to let go yet. His hatred paralysed him and for a moment he stopped breathing.

And then he was much younger, and she was a towering mass in a blue print dress against which he butted without effect, knee high and weak in a child anger that had no possibility of resolution and ended as it always did by tearing away, shrieking, looking blindly through tears for the corn fields where all the TV kids ran to escape the adult world. Then he was back.

"And what am I, you demented whore? Something you fucked up so totally it never had a chance to make it into the world. Jesus, it's as much as I can do to walk down the street."

He vomited tiredly on to the table, bracing himself against it with locked arms. The Hagbeast laughed softly and clumped across the room to stand over Dog on the shit tray.

"The time wasn't wasted then."

She lifted her skirt and pissed on the whimpering animal.

Matthew Stokoe

In the drifting monochrome wash of the TV, Dog's coat looked dark and oiled. The hair parted in a rolling wave as Steven dragged a towel back and forth across it, exposing a narrow moving line of white skin and the occasional cluster of fleas. The stink of the Hagbeast's piss burned acquisitively through the dead air, snouting out strongholds in the dampspore that blackened the corners of the bedroom, planning to linger. Dog grunted happily under the attention but its eyes held the sad light of betrayal that surfaced with each of the Hagbeast's cruelties Steven failed to protect it from.

Chapter Ten

He had been to the fourth floor before. In the endless years of his growing it had been part of a route that lead to a temporary escape from the mad bludgeonings of his mother. Up the stairs that were always unlit and creaked fear into a young boy's legs, along a landing so thick with its own isolation that the shadows must, absolutely must, hide something hideous and fanged that drooled for the blood of a child, to a ladder at the end that you climbed to a square of meshed glass, then out on to the roof, panting and pushing the skylight open and gulping down the gritty air of the city that seemed back then to expand in your lungs and float you up and out into a world not shared by monsters or mother.

Beyond an immediate ring of desolation, lights glittered and flashed colours out across the world. And the colours were so significant then – each neon shade tugged at him with its promise of a different way to live, each glowing curl of tube-glass was an entire world that would close around you if you stared at it long enough and carry you off on warm purple nights to a place where there was

Matthew Stokoe

music and people laughed.

To stand by the railing at the edge of the roof, kicking at loose bricks and dreaming of moving out into those lights was payment enough for the shrieks and beatings that inevitably greeted his return to the flat.

But time clawed its way across the lights and they paled. They took on a new meaning that brought no gladness to Steven's heart. Where they had once been the fuel of dreams they now became a cankerous reminder that those dreams had not come true. So Steven stopped climbing the ladder at night and began to search instead the less fickle TV screen for ways to the worlds he had seen from the roof.

Now the fourth floor was different. 40 watts burnt over a dusty grey carpet runner and the sepia light showed Steven only a duplicate of his own landing. The haunted infinite darkness he had imagined as a child had been exposed by a weak light bulb and the passing of time as a deception. It was no longer the mystical, horrored passage to dreamtime that had so attracted him in those early years.

But standing there, silently gathering his courage outside Lucy's flat, he could not help hoping that it might again become a section on some road to happiness. Not the real happiness that TV so accurately threw across the bareness of his bedroom – he could not hope for that – but an approximation of this ideal, a stockaded copying, built with the only materials to hand, within which his loneliness could be shallowly buried.

Lucy opened the door, then drifted back and collapsed on a couch. Steven followed her in and sat at one end. The room looked like it had been picked up and shaken. A thousand small objects lay scattered over any surface that would hold them. Some of them were clothes and containers of food, but many were shiny steel and surgical in nature. Small lamps shone yellow in corners and a video played an abdominal operation – close-ups of blood on green medical cloth, tight angles on smeared rubber gloves probing inside a human, low volume technical commentary.

"They sell these to people who want to be doctors, to teach them. But I don't think they look for the right things."

Lucy talked without taking her eyes off the screen. The surgeons manipulated organs and she started to shout, jerking forward, squinting.

"Look! Did you see that, when he lifted the liver?"

"What?"

Lucy rewound with a remote.

"There was something under it. Didn't you see? It was black and shiny. Look."

The tape played again and the black thing under the liver was only a cavity filled with blood.

"Shit."

Lucy slumped back, but she didn't stop watching.

"Why don't they show it. One day it'll be there. They'll forget to hide it and I'll know just where it is."

Then remembering, turning to face Steven.

"Did you look in the cows?"

"There wasn't anything."

Lucy's face set.

"You didn't look."

"I did. I looked. I moved everything around and I couldn't see anything except guts."

"Did you look inside the organs?"

"Some of them."

"What about the intestines?"

"How could I? They're all clogged up with shit."

Lucy was angry.

"It could have been in there. You should have checked."

"There wasn't anything there."

Lucy sucked her teeth in disgust and stopped the video. Steven was worried, he needed to make a connection. This room and its disarray, this girl with tits under her T-shirt and her legs sprawled apart was the closest he was going to get to a wife and a ranch in the country. He tried to sound sympathetic.

"How do you know there's anything there at all?"

"Because I know how much pus my body churns out. I've measured my shit and my piss and my snot and all the other slime that comes out of me. And it doesn't add up to what being alive pumps into me every fucking day."

"If you're so sure how come you have to find it in cows or see it on a video?"

"Because if I know exactly what it looks like and exactly where it is I can find it in me and cut it out."

Lucy pushed herself up from the couch and walked over to something on a table that looked like a computer. She fiddled with the console and picked up a thin black flexible rod that was connected to it by a length of wire.

"Help me look?"

She pressed a switch and the monitor came to life, showing an unfocused disc of shadows and light that shifted as she moved the black cane through the air. Steven could see a bright light at its tip.

"It's an endoscope. It'll show if there's anything in my colon, but I need you to help me put it in."

"Sure."

Lucy pulled off her tights and bent forward, bracing herself against the table, in front of the monitor. Steven smelt shit as he worked lubricant into her arse. Her ring was tight like Dog's. He couldn't tell if there was anything sexual in it for her, but they'd got intimate awfully fast and pictures of a future he never expected to see were scrolling up into the present.

"Push it in slowly. I had a shit before, so it should be clear."

Steven eased the probe in. The light glowed through her arse briefly, then it went deeper and the disc on the screen focused. A close tunnel of yellowish gut bulged fatly in from the edge of the picture, its centre shadowed, beyond the reach of the light. The probe slid smoothly for a few inches then hit a bend. A head-on view of colon wall filled the disc, so brightly illuminated that the dark veins beneath the surface were visible. Lucy tensed and sucked air.

"Sorry."

"You can steer it with those handles."

Where the probe became a solid haft there were two steel loops like the rings on an old-fashioned syringe. By pulling back on one or the other of these, Steven found he could twist the head of the probe enough to guide it round the curve and into the next section of bowel. Here the gut lining was more corrugated and the folds were crusted with hard deposits of shit.

Lucy made a noise.

"Jesus, that's disgusting. Even dumping doesn't get you clean."

She shook her head sadly.

"My parents used to tell me to be happy. What a fucking joke. How can there be any happiness with filth like that rotting away inside? You've got to be clean to be happy. Go on, push it further."

"Isn't that what you're looking for?"

Steven was more interested in watching the probe disappear into her hole, than in the image on the screen.

"Fuck no, that stuff just comes from food. The real poison comes out of your head. All your fuck-ups and sadnesses and fears drop down like some sort of brainshit into your guts and build up there. That's what really fucks you up. I told you before."

Lucy had over a foot and a half of the probe in her now and her teeth were clenching against pain. Steven bent the endoscope through a particularly tricky twist of colon and slid it forward another few inches. The blunt head of a turd blocked the way, like an animal in its lair.

"Jesus fuck, not more already."

"Do you want me to push it through?"

"Can't. It'll smear the lens and you won't see anything. Leave it there, I want to look at it."

Steven let go of the probe carefully and leaned back to get a better view of the rod sticking out of her. He stroked the skin around her cunt, she didn't turn.

"We can look in you after, if you want."

"Huh?"

"You'll have poison inside you too. The only way to get normal is to find it and cut it out."

He slid his middle finger into her cunt, she was wet and she pushed against him.

"It's too late for normal."

He got his cock out and stuck it in her. He had to bend the probe out of the way and the picture shifted slightly.

"Keep it where it was."

She sounded urgent so he twisted the thing until the shit was centre screen again, snouting blindly into the glare of the endoscope.

The picture vibrated as Steven pumped, but Lucy was locked in on it too tightly to complain. He watched his dick plough between red flaps of skin and thought he could feel the hard line of the probe pressing against him through a layer of internal meat. Near the end Lucy started to moan.

When it was over he pulled the probe out of her arse and she cleaned the streaks of shit and intestinal mucus off it with a handful of tissue. She held the stained paper to her nose.

"Fuck, it stinks."

Chapter Eleven

Steven thought it might happen straight away, that the fucking might magically bring into existence around him the world of his TV dreaming. But Lucy lay down on the floor, in a cold draught from an open window, and fell asleep without speaking to him. So he went back to his room and the TV and Dog in the corner with a dog smile so happy to see him, and lay naked under his torn blanket, lifting and lowering it, puffing out the fish stink of his recent sex in warm gusts of memory.

He was not worried at the delay. She would be what he wanted, Lucy upstairs. He knew it. There might be more steps to take, but she would be the mother, the lover, the hook on which to hang his plagiarised blueprint for living.

There had been no love there, upstairs tonight, but it would come – Lucy would force it into being. She had no choice. She would never find her black lumps of poison or cut them out, and like Steven she would never be part of the world. In time, when she realised this, she would need someone to cling to, someone to absorb and deaden the impacting horror of her sentence. And to justify this dependence she would have to call it love.

The Hagbeast would permit no such joining, of course. She would move swiftly to destroy any source of affection, any avenue of hope, that threatened her tyranny.

And so she must not find out.

But that was impossible. How could he hide from her a growing involvement with Lucy when she tracked the slightest of his movements with every sense she possessed? He was transparent to her and sooner or later she would know, despite any camouflage he might erect, that he was directing himself to more than his daily struggle against her.

She would know. She would home in and ruin his dream before he had a chance to use Lucy to make it real. She would expel him from the flat or she would kill him. There could be no middle ground.

Here, now, with Lucy's cunt scum crusty on his dick, with the raw materials of his envisioned satellite world at last close enough to reach, the inference was obvious. And it did not surprise Steven that he found little horror in its contemplation – he had suffered too long.

Steven did not sleep.

How could it be done?

What would it feel like to kill, to actually extinguish the pile of meat that had shitted him into existence? If she had been a mother like mothers were meant to be, then he supposed it would be impossible. Or if possible, that it would trail such jellyfish tentacles of remorse his eyes would be forever clouded with the final stinging vision of thick white foam boiling past her swollen tongue and out over his wrists.

But she had never worn a blue checked apron or baked sweet pies in a kitchen where the warm air made her cheeks rosy, never reached down with floured hands to lift him up on to the table and kiss his face and laugh at his giggles with her eyes so bright he thought he would never see anything else again, or want to, never shown him things or let him press the dough before she wrapped him up in herself and carried him off to bed. And because this was so, he knew the act would not bleed forward in time to harry him in small-hour awakenings. It would stop when she stopped.

The killing would bring him relief, but its doing would not be easy. He could imagine himself, head back and howling, in a suffusing glory of murder, gouting semen across her naked shoulders as he hauled back on her head and snapped her neck. But reality would be different. Reality would be a frightened rush to the finish with no time to linger over details, a headlong plunge to get it over with before his courage gave out, before a lifetime of conditioning reared up and robbed his arms of strength.

Steven squirmed in his bed. He had to do it, there was no other way. But in twenty five years he had not lifted a hand against her, and thoughts of starting now with the ultimate hand-raising made him frightened enough to puke. His body felt boned and unequal to the task.

Much better to find some way less direct. Killing without the necessity of active throttling, stabbing, beating participation. She was old and immensely overweight, the systems of her body degenerating under an onslaught of filth and the mordant ravages of age. There had to be a way to place a final, terminal strain on them. An iceberg method that kept the bulk of its guilt and purpose hidden from sight.

Steven watched the shifting reflections of streetlight on his ceiling until dawn.

Chapter Twelve

"You were out last night and you didn't tell me."

The Hagbeast ladled an oily sludge into his breakfast bowl. Her eyes flicked blankly across the table and into the corners of the room, a false front running interference for her words.

"You know Mama needs to know where you are all the time."

"Why?"

"So I can be sure you're eating right."

"You shouldn't have pissed on Dog."

"I should have pissed down its throat and drowned it. Where did you go?"

"The roof."

She chuckled, it made her neck shake.

"You moron. Staring at the people out there won't change anything. You can't be like them, don't you know that? You're part of me, you little fuck, part of this place, and you'll die here."

Steven tipped whatever it was she had served him out on to the table and threw the bowl across the room. He

---

didn't bother to stand.

"And when I do it'll be a long time after you and I'll have someone to love me when I go."

The Hagbeast snorted into the greasy early morning air.

"Who's going to love you, Steven? I used to live out there before you infected my cunt. I know what they like and what they love. And it isn't you. Hear me, you piece of shit? It isn't you."

She spat on the floor and caught her breath.

"Clean up that mess, you fucker, and eat."

Steven didn't move. He looked into those empty eyes and decided it was time to test himself.

"I know what you're doing with this food."

The Hagbeast's face went dark with blood and she shouted each word distinctly.

"I am not trying to poison you."

"Yes you are."

"I've told you before, Steven, I eat what you eat. How can it be poison?"

"Because it is. I can feel it in me."

"For the last time, cunt, it's only food. Now eat it."

"It's shit."

"If I eat it, you will as well."

"Not any more. From now on I'm going to make the food."

"What?"

The Beast lurched upright, slavering and working her mouth incredulously. Fat slewed about her frame under the sudden acceleration. She planted her fists on the table and roared.

"No!"

The stink from her throat wrapped itself around Steven's head. He stood up, breathed it in, drew back his

---

arm... and hit her. A single short hook to the side of the head. He felt the impact travel through his bones, the sandpaper crunch of his knuckles against the coarse skin of her face. For one wild moment he wanted to keep on hitting until she was a bleeding sack of shit, draped shapeless over the back of her chair. But he couldn't do it. Instead he watched a white smear of disbelief shade out from the red mark on the side of her head.

She looked at him through eyes veiled with the calculation of shifting power balances. Her features held no trace of pain, only a drenching hate that boiled with the reassessment of options.

Steven held her gaze, but it was a war. The hard seconds thudded into him, working on his knees and stomach, searching, weakening, all the time getting closer to finding a path to that soft interior where reassertion of the dominances scattered by his blow might be possible.

It was time to go. Her scrutiny threatened the glory he felt burning about him like the cold fire in some picture of God. This first, small act of defiance was too valuable to be risked here in the flat light of the kitchen. It must be gathered in, protected, allowed to grow and to extend into time, raising structures in its slipstream that would shelter him in the future.

He put his head close to hers and said into her face, "I make the food and you eat it."

He left the room as she started to shriek.

"Fuck you, you fucking moron. I know what you're doing. Anything you can make, I can eat. My guts held you for nine months, you can't get worse than that. You think you can beat your mother? We'll see. We'll see about that, you dogshit."

Her ranting followed Steven down the stairs like garbage tipped from a pail.

Chapter Thirteen

On the line that morning a cow got loose, somehow slipped from a grabber before the slaughtermen put the bolt in its head, and came clattering into the process hall, half slipping on blood, scattering men, ramming the inverted dead bodies of its brothers. Looking for an escape from cow hell. But its terror must have made it blind and it ended by slamming its soft nose against a ventilation grille until Cripps came over and blew its brains out with a shotgun.

The ruthless efficiency of the killing took Steven's breath away. Cripps moved without doubt or hesitation. He did not consider the phases of his attack, he simply saw a problem and removed it in a flawless, perfectly economic stream of action.

If Steven possessed such clarity, such sureness of purpose, ridding himself of the Hagbeast would pose no greater problem than crushing an insect. At breakfast he had decided to poison her as she was poisoning him, but now considerations of practicality had begun to cool the fire that had earlier singed his veins with the ecstasy of confrontation.

Could he force himself through it?

Would she really eat whatever he gave her?

And if it killed her, would his own body be strong enough to survive?

The resolve of just a few hours ago was becoming infested with the worms of doubt.

Cripps had spoken of mastering the self, of releasing a potential for action that benefited no-one but the individual concerned.... Of selfish epiphanies in blood. And Steven wondered, as he watched him carry his shotgun back to the slaughter room, if there might not be some crutch beyond those plastic strips that could support him through the killing of the Hagbeast.

The afternoon shift was half through when Cripps appeared at his side and took him away from the grinder.

"You look ready, boy. I've seen you watching the slaughter room and I know what you've been thinking – 'Is he right? Is there something in there for me?' Well, I am right, boy. The slaughter room gives up its secrets to any man with the cock to ask. Are you asking? Have you got the cock for it, boy? Have you?"

Inside the slaughter room death was in full swing.

The place was a storm of bawling cows and goading, muscular men working with fierce precision. These men moved as Cripps had during the shotgun execution – without weakness, without even the thought that they might position a hand or a foot unsurely as they punched and kicked and prodded the animals with stubby electric lances along the alleys that lead to a final bondage of pneumatic presses. Some were stripped to the waist, all were streaked with blood and wet cow shit. They sweated and wrestled cows into position, faces creased in tight grins of effort, taking pleasure in their own strength, calling to each other

Matthew Stokoe

over the din, directing, pointing, clapping hands like it was all a play in some bloody contact sport.

Some of the cows in the alleys bucked against the rails, trying to turn and plough back into the reassuring brown and white and black cowmass, rearing up and scrabbling at steel and brick with slippery hooves, eyes white all the way around, nostrils wide, snorting in as much air as they could hold, knowing that its taste would soon be lost forever and trying to imprint it on some soul memory so it could be remembered after death, shaken out like a table cloth and searched for meaning. Others trotted madly in a straight line, refusing to see the swinging V of the grabber in their path, running only for the blur of white light from the process hall that maybe looked like freedom. Like moths.

On the platforms by the grabbers slaughtermen worked the boltguns on their counter-weighted chains.... Swing smoothly forward over the rails, nudge the muzzle into the soft hollow behind the ear, look at the cow and wait to make sure it knows what you're going to do, then pull back on the trigger and send a four inch hardened-steel bolt through skull and into brain, swing the gun away with the bolt already retracted by recoil and watch shit squirt out of one end and blood out of the other.

Where the room had been empty and awkward that other lunchtime it was now hot and bent to its purpose, seamlessly fusing the action it housed into an organic whole where airborne blood and shit and beasts and brick walls and steel girders became one in a designed and streamlined operation.

Steven watched it all and wondered what he was supposed to feel. It was obvious that these men moved within the flow of some connecting and energising force. They shared a motive confidence that made them even more intensely alive than the others out here in the world. The sight of them roused him to envy, but the staggering deaths of the cattle as they collapsed against the grabbers did nothing to stir a corresponding sternness of self-direction in his own breast.

"Majesty, boy. The death of animals and the rebirth of men. You can feel it, can't you? There is glory in this room. Look at them. Many were like you before they learned the secret that killing holds. Timid. Yes, boy, timid, but with the cock to push themselves beyond what they thought they could endure. They didn't know what they would find, but they went looking anyway. And when they confronted their own uncertainty, when they crossed to the place weaker men had forbidden them to enter, they found a strength greater than they ever dreamt existed. Come close and watch."

Cripps lead Steven on to the low platform beside one of the grabbers and held him tightly about the waist while they watched the slaughterman work. A cow was driven between the iron jaws of the grabber and Cripps whispered harshly into Steven's ear.

"See how it comes, so full of life – eyes seeing, mind thinking. Life! Prized above all other things. Touch it, feel it breathe."

Steven leaned over the guard rail and put his hand on the cow's back. The slaughterman watched, ready with the boltgun but waiting. The cow felt solid and warm.

"Keep your hand there."

Cripps nodded and the slaughterman put his gun close against the straining bovine head. Steven felt no particular affection for the cow, but the fly-shooing tremors that jerked in waves across the animal's hide shook his arm and jarred loose within him broken-glass splinters of panic. He was about to feel something die.

When the gun went off the cow threw itself forward and collapsed like an enormous rubber toy, pumping steaming liquid shit down the inside of its thighs.... Off into cow darkness.

Steven snatched his hand away and looked quickly to see if it had absorbed the mark of death, some dark contagion that might multiply beneath the skin and come searching for him.

There was no mark, but the shock of the killing sent small blurts of bile into his throat. Cripps was laughing and pressing a hard-on against the side of his leg.

"Did you feel it, boy? Did you feel it just... stop? It's like a switch, isn't it?"

"Yes."

"You must be aching to try it yourself."

How far did he have to go for this magical awakening, this unleashing of strength that Cripps talked about? He was already flecked with blood and shit. He had seen the bolt punch into cow head, rip away a circle of hide and bone and slam deep into cow brain. He had smelt the fear and the last rush of breath and the emptying bowels and the wet-newspaper mustiness of the inside of the cow's skull. And there had only been horror at the ease of it all, the sickening backwards flip of approaching fugue – not the sunrise of a new way to live. But might the secret be waiting a little further on, standing elegant and incurious beyond a bleeding threshold of flayed beef, needing only a little extra letting-go to be caught?

The boltgun swung heavy and smooth on its supporting chain, its butt warm from the slaughterman's grip. The chipped grey enamel of its surface was brightly caught in the purple-white net of light that fell from the halogen spot above the grabber. Steven could see the thickness of the paint, and the minute shadow this thickness cast on the scratched patches of bare metal. The slaughterman helped him guide it to a new cow, his hands were hard and crusted with blood.

Everything narrowed down. Steven saw the muzzle of the gun and a tight oval of light brown hide immediately beyond it. There was nothing else. The activity of the slaughter room rolled away like stage scenery into some distant other world and he was alone with white noise hissing in his ears.

In this blurring, roaring cocoon he felt the weight of the gun, and he felt Cripps against his back, arms circling to the front of his pants, unzipping, pulling down.

Then Cripps was in him, pounding at his arse, whispering encouragements he couldn't understand but which filled his head with a mounting pressure, and the gun felt more real than anything he had ever touched. He had both hands on it and Cripps' breath hot on his neck, and he knew the cow was pissing on the floor with the agony of the stretching seconds and then... time... stopped.... Until something sucked away every sound that had ever been made and the world zeroed to aching curled fingers and the shadow of the gun on the cow's skull and he pulled the trigger as Cripps shrieked somewhere a long way off and sprayed seed into his arse.

Slumped over the guard-rail. The end of shift horn sounded dimly out in the process hall. Steven felt the flaccid length of Cripps' withdrawal and opened his eyes to the twitching, fallen carcass of the cow and its dark collar of blood. Strong arms pulled him upright, tearing the white crepe of his half faint, shunting him back to the din and the killing and the mad, channelled exertion of the slaughter room.

"That's it, boy, breathe deep, breathe deep."

Cripps' voice was gentle as he led him to the

---

observation platform overlooking the slaughter floor.

"Lie down."

Steven curled himself on the concrete, looking down on the men who still appeared to be working despite the end of the shift. Cripps sat beside him, touching his shoulder.

"The nausea is normal, it will pass. Your body is reacting to change. You have killed, you have started to learn."

The work on the slaughter floor had changed. The men stood close to a single cow held helpless in a grabber, passing round an instrument like an apple corer. Each in his turn used the serrated steel circle to cut a hole in the side of the animal. Blood ran down the curve of its belly and pooled between its feet, but it remained conscious and standing, bellowing its humiliation to unseen cow gods who couldn't be bothered to answer.

The room went dark at the edges and Steven felt again a tightening of vision that excluded everything but the spot-lit cow and the crowding men. Gummy had appeared from somewhere and was bent close to the animal's hindquarters.

When all the holes were cut the slaughtermen pulled out thick, hard cocks and stuffed them into the wounds. Steven watched buttocks clench. Three men on either side, linking arms over the back of the cow to counter-weight their thrusts.

"Do you see, boy, that you still have some way to go? Your killing was a stumbling first step. These men have learnt to run."

"Gummy...?"

Steven's lips felt numb, it was an effort to speak.

Cripps laughed quietly and sneered.

"No, not Gummy. We give him this as charity."

Steven's eyes were heavy but he kept watching. Down on the floor, while the men rammed in and the cow screamed, Gummy, his open mouth sucking the animal's arse, slid a cattle prod into its cunt and triggered the electric charge. The cow's rear legs lifted off the ground and Gummy fell backwards under a blast of shit, vomiting in rapture.

The slaughtermen hung on and moved faster, blood on thighs and stomachs, howling through corded necks until one of them fired a boltgun and made the beast close like a fist and all six of them shot seed into the torn, dying guts that had hoped one day to swell with the weight of a calf.

Steven's eyes closed.

Matthew Stokoe

Chapter Fourteen

At home. In the kitchen Steven played his mind against itself, diverting it from the slaughter room obscenity with small domestic actions. And then, when the deception of these actions became too obvious, ricochetting back into curtains of blood and streams of semen splashing from jagged cow-hide holes.

He drifted in the kitchen, blank faced, picking up plates and wiping them, putting them down, wiping them again, polishing cutlery against the side of his leg. Somewhere at the back of the flat the Hagbeast made dim shunting sounds as she moved about, but Steven didn't hear them.

The killing of the afternoon was stored inside him, weighted by the heavier, following torture, but he was afraid to examine it, afraid to search for its effects. That part of his brain was temporarily locked.

And he was afraid of what he was going to do now, with these plates and forks and spoons. This was the beginning he had wished for but never expected to see. Tonight the Beast would eat the first of the meals that would send her down to hell. But if he failed? If he hesitated or was weak? Then she would rise like a gorgon and split him open.

He had planned, on the bus the morning before Cripps' horrorshow, to use some disguised ingredient subtle enough to escape detection and of a borderline virulence that would eventually destroy her, but allow him, stanchioned by youth, to recover.

But now...? But now...?

As he squatted in front of the cupboard under the sink, staring at ancient and unused bottles of disinfectant, bleach and drain-cleaner, trying to choose between them, he felt a sudden wild boldness flood his guts. Subtlety was pointless. She would eat whatever he did. She had to, her hate for him would not allow her to refuse the challenge.

He took two empty plates into the bathroom.

It was dark when the Hagbeast galleoned into the kitchen. The bare overhead bulb cut hard shadows into the sheets of newspaper tented over the plates on the table. Steven was seated and waiting.

"So, we have a new cook. What did you cook, Steven? Uncover it. Let's see if you can match your mother."

Steven drew away the paper and watched the tight compression of her smile, the narrowing of her eyes. On the plates, equally portioned, two curving lengths of shit lay dark against veined china.

"It won't work, Steven. Do you think this is so alien to my system?"

It won't work.... Steven went cold. She knew what he was trying to do!

But she was pulling her plate towards her, pressing her fork into the softness of the stool, lifting a piece to her mouth. Her eyes in their mean folds of fat held his, and for

Matthew Stokoe

a second the stink of shit absorbed time. Between them space emptied of all the mists that usually swirled there and Steven saw how well she understood him.

Then she moved and the stink was just stink again and Steven had to carry on, whatever she knew. He saw thin fibres and lumps of still recognisable food poking from the broken end of the shit and prayed that her destruction would be swift.

The Hagbeast waited for him to eat first. He put a section of the shit into his mouth. It rubbed his lips and the chocolate-smear drag of its entry made him shudder. He could not immediately bring his teeth together and the turd lay acridly in the hollow of his tongue, forcing its thick, boggy smell up behind his nose and into his head, cinching his stomach in a rapid serial spasm that threatened to send bile squirting from his nostrils. He forced himself to bite down and chew quickly, but speed didn't reduce the appalling foulness of the taste.

The shit was gritty against the roof of his mouth and made crunching sounds with his teeth. It worked itself into a clogging paste that built up under his tongue and inside his cheeks, so stiff he had to use his finger to hook it out. He felt he was drowning in the anus of some dysentery-struck mammal, vistas of the world made shit opened before him. Then, at last, a small amount of vomit punched through his locked throat, into his mouth, and mercifully allowed him to swallow.

He bent forward and gripped the legs of the table, screwing his eyes shut. Thin brown liquid ran from the corners of his clamped mouth and he jerked quickly on his chair, up and down, fighting his stomach, willing it to accept the returning waste.

Somehow he kept it down and when he looked at the Hagbeast again her smirk had faded. It was her turn.

Shit in her mouth made her twist her head in a spastic half circle and pump her neck into a tightly stretched red bag, like some obscene mating bird.

The force of her first retch blew snot into the air, but it didn't part her lips. She lurched against the table, then steadied herself with weak arms while her belly shook. Bunching jaw muscles showed through the loose skin of her jowls and the sound of grinding teeth made Steven press his thighs together. How she must be damaging herself to compete with him.

Then she couldn't hold it any longer and puked on to her plate in a screeching explosive torrent that spattered the front of Steven's shirt. She heaved a few more times, until it came up dry, then sat, arms rigid to the edge of the table, shivering and silent, drawing breath. Steven felt dismay creep into his already churning guts. If the Hagbeast could not master a plate of shit how could he fill her with enough poison to kill her? He saw his plans crumbling and was about to speak some desperate goading remark when her arms relaxed and she began to function again. She cut a piece of shit with the edge of her fork, speared it, put it in her mouth and swallowed. Her movements were deliberate, machine like. She cut another piece of shit and ate it. Small tremors rippled across her breasts and shoulders, but they did not touch her throat. She looked at him and smiled ingenuously.

"Steven, I can't keep eating without you."

He slid his fork into the thing on his plate, thankful that it had escaped most of her vomit – her own plate dripped, the shit swam in it – and entered again the body rebellion of his first mouthful, and kept forcing it in.

"How is it?"

He did not look at her as he spoke.

"It smells like you birth. I didn't expect this from

you, Steven. You've started a game with your darling mother, haven't you? Those years in your room with that fucking mongrel and your precious TV, doing nothing but wanking and picking pus out of your face, and you think you can just crawl out and wipe off the slime? Just reach into your box of dreams and slip one on like a coat? You sorry fuck, you're not strong enough to do it."

"I think I'm getting stronger, Mama."

The Hagbeast laughed and opened her mouth in mock surprise. Steven saw bits of shit stuck to her teeth.

"Strong? You were born a runt and you haven't changed. How strong are you getting? Come on, show me."

She finished the last nugget of shit and smashed her plate against the table.

"Get out of that chair and stand up. Mama wants to see how strong you are."

Her bellow hit the dead walls of the kitchen and came back at Steven in a rolling chain of thuds, each one pushing him further upright, until he stood, arms limp at his sides, waiting for the coming humiliation. God, if he could be like Cripps for just one minute....

The Hagbeast moved close to him and their breaths combined in a sluggish cloud of shit and saliva. She was too close, he shut his eyes. He felt her fat fingers undressing him. His cells screamed, but his arms were too weak to fling her from him. Too weak to force her mouth apart until her jaws snapped, too weak to yank her head down so sharply that the spine broke a few vertebrae from the skull and stuck out into the air through the skin at the back of her neck. Too weak to enact a thousand killings wished a thousand times. He had spoken too soon.

He was naked.

"Look, Steven." She hit his face. "Look at yourself."

Steven looked down and saw what had always been

---

there – soft white skin over bones, ribs, dick hanging.

She laughed, prodding his chest and stomach, lifting his balls to look underneath.

"I don't see it, Steven. Where is this strength of yours?"

He stood mute. She was too powerful for him to survive direct and active confrontation.

The Hagbeast reached down and pulled her dress over her head. She wore nothing underneath and the sharpness of her crotch burnt his throat.

"Are you as strong as this?"

She slapped her dimpled saddlebag hips, ran her hands over rolls of hard fat stacked from groin to breast. Steven looked at her matted grey cunt and the blood sticking in clots to the insides of her thighs.

"Look at this mountain of flesh, Steven. Throw yourself against it. Have you ever calculated its weight? This is strength, you whining bucket of piss. This is what you must measure yourself against. It stands between you and everything you want and you'll never get past it."

Steven knew she was wrong and he wanted to spit it in her face. Lucy was going to open up like a tunnel and he would crawl through her into a world impossible for the Hagbeast to touch. But it was too early yet to strut this before Mama, she could still destroy it at a stroke. So he stayed quiet through her ranting.

Later, in his room, the shit in his belly made him sick and he lay curled around Dog on the floor by the bed. Dog licked sweat from his master's forehead and whimpered at his shiverings. Steven felt the animal's nuzzlings through the gauze of his pain and dreamt he was somewhere underground with the velvet lips of a cow against his neck. In his fever he merged with it, knowing its thoughts, its fears and the timeless species-desire for a place where men never came.

At dawn he was able to rise, pale and drained, and Dog yelped with joy and gave thanks to Dog God that there was still something left to love.

In the hall, as he left the flat, splashes of the Hagbeast's vomit bloomed on the floor from the kitchen to her room, like flowers of hope. Steven felt good when he saw them.

Chapter Fifteen

The door was unlocked and Lucy was up, so Steven walked in and stood behind her as she sat bent over a table. He kissed the back of her neck and looked over her shoulder. Pinned to a wooden block, a lab-bred rat lay on its back, alive, guts open to Lucy's probing fingers. The rodent's stretched eyes darted uselessly through a limited range of vision, searching for some way to escape the pain.

Lucy gave up with her fingers, took a scalpel from a clutter of sharp surgical instruments at her side, and began to remove, one by one, the exposed organs. She held each one up to the light and inspected it, then cut it into pieces on the wooden block.

Steven kissed her hair as she worked, running scenes of the future when she would tend to him with the same devotion she now reserved for the small rat organs. He would lie in a large bed under her kisses and plans for life would flutter down about them like rose petals.

When the rat was empty Lucy dropped her scalpel and leaned back against him, too disgusted to support her own weight. He could feel desperation radiate from her.

Matthew Stokoe

She put his hands on her breasts, but this was too little protection. She stood for him to hold her and closed her eyes and pressed loose fists to her chin like a baby sleeping.

Steven saw her night's work on the floor – a pile of hollowed rats and a plastic bucket of guts – and knew that love was coming quickly. Her search for something to cut out of herself was getting frantic.

They fucked in a cold room hung with photographs of surgically opened bodies. Steven looked at them while he pumped. The light in the shots was hard and the exposed organs gleamed under it – dark kidneys and livers and hearts, paler stomachs and bladders, all of them floating in cavities of blood like the makings of some hideous stew. In one of the pictures the incision was stretched so far open it showed a cross-section of abdominal wall. The striations of meat and fat made it look like a piece of bacon.

Afterwards the concrete-dusted light of morning fell across them. They stared at the ceiling and Lucy's cunt leaked the ichor of their beginning into dead sheets. Steven thought of the slaughter room, the recoil of the boltgun, blood and come sliding down the sides of a punctured cow, Cripps in his arse. The act of killing.

"I killed a cow yesterday."

"Were you trying to look inside it?"

"The foreman said it would change me."

Lucy laughed softly, sliding towards sleep.

"It isn't that easy."

The sun hauled its broken-backed way higher into the aching slum air, turning the windows dirty yellow. Was he different from yesterday? The Hagbeast had destroyed him at dinner as easily as she always did. Where was the muscle-charging certainty of action Cripps promised? The slaughter room cow-killing had overwhelmed him to the point of unconsciousness and he expected something in return. But all he felt now as he thought of it was a lingering revulsion at its bloodiness.

It got late and he went downstairs to wash the sardine stink off his dick and have a shit. His arse was sore and all he could force out were small dark pellets that stung his ring and lay heavily under the water like a handful of stones.

The Hagbeast wasn't up and Steven turned circles in the strange freedom of the kitchen, gathering armfuls of joy at this foretaste of her absence. He drank water and felt it clean him. Then he left for the plant.

Chapter Sixteen

The bus trapped sunlight that morning, the air in the aisle was hazed with it, and through arabesques of cigarette smoke and the chaos intricacies of floating dust the other passengers seemed less than they had been. Not quite the gods of yesterday.

Steven wondered at the lightness he felt, paranoiacally fretting he might burst into laughter, right here in front of everyone on the bus, at these first gossamer strokings of happiness – so unused was he to their touch. What brought them? The time with Lucy? The Hagbeast's first plateful of shit? Or could this elation, this feeling of possibility be a delayed gift from a dead cow? He flexed his arms, twisting the muscles to see if he was stronger. He couldn't tell.

Half an hour later, the drifting, window gazing euphoria of the bus journey evaporated as he entered the process hall. Here things were real again. The weight of the boltgun and the spurts of blood were no longer smooth-edged pre-fugue memories, but intense and unavoidable occurrences that stuck sharp red fingers of recognition into his head and refused to be ignored.

He walked past the other men with his eyes on the floor, ashamed they might see the mark of the slaughter room on him and know the intimacy of his experience there. He sat by himself at the grinder, staring at the scoured steel work surface, dazzling himself with the million curving scratches that caught light and bent it into a bright flat tangle.

The flow of meat started with the horn and time passed in chunks of bleeding beef. Steven worked hard and tried not to think, because when he did he got confused. He didn't understand what had happened in the slaughter room. It had frightened him.... And yet there had been that flash of happiness on the bus. Now he was frightened again – of the blood and the cutting of holes into cows and the mad, wantonly exposed selfishness of the slaughtermen, and of not knowing what all this had done to him.

When he heard the voice behind him he froze, thinking it was Cripps. But there was too much smoothness to it, too much humid depth for it to belong to the striding, blood-bathed foreman.

The voice called his name again and it came through a lot of throat. Steven twisted quickly on his stool.

Just a white wall and, down near the floor, the ventilation grille. Then movement behind the grille and Steven was on his knees, peering through it, pressing his face against the mesh. In there, in the shadows beyond the spill of light from the hall, the outline of an anvil-shaped head swayed gently. Two eyes blinked limpidly, insolent in their slowness. A dark mass moved forward into the light.

"That Cripps man is going to fuck you up, dude."

It was a cow. Most of the body was below floor level but Steven could tell it was a full grown animal. A sienna Guernsey. He looked closely at the flawless sandy curves of forehead and cheek, at the chocolate darkening of the mouth and nostrils, at the badger rings around the eyes. For an absurd second he thought that if he looked hard enough at it the thing might phase back into his head and disappear.

But it was real and it stayed.

"What...?"

"Yeah, I'm a cow, man. Touch me."

Steven stuck his fingers through the grille. The cow was a cow, warm and solid.

"Can you handle it?"

Steven nodded, but it didn't mean anything one way or the other.

"Good. Listen, man, you keep going to the slaughter room with Cripps, you're gonna to get fucked over. You think it'll help, but it won't. You got sick last time, learn from that."

"How do you know?"

"Ah, man, we're always watching. And we know Cripps. He's been here forever and this ain't the first time it's happened. He told you the slaughtermen weren't like other men, right? He talked about power and freeing yourself to take whatever you want. And you thought 'Shit, that's just what I need. He's right, look how different those guys are'."

"I didn't know what to think."

"Yeah, but you wanted it, didn't you?"

"Who wouldn't?"

"Sure. But can't you see it's a load of shit? Course those guys look different, but it ain't because they've gotten to be better people. Shit, they spend all day chopping us up and raping us, it'd be fucking weird if they didn't look a little different. But it ain't magic like Cripps says, no fucking way. What it is, man, is a way to stop yourself feeling, and you need your fucking head read if you think that's the way to go."

Steven sat back on his haunches, head gridlocked by a stream of cow words he didn't want to hear. He wanted the strength Cripps promised, he wanted to change himself into someone on TV, someone who had the guts to get rid of the Hagbeast and build himself a life.

"You might be wrong. How can a cow know what changes a man?"

The cow stamped and rolled its shoulders.

"Hey, fuck you, man. You think we're stupid? We watch those guys outside this place and they ain't the supermen they think they are, believe me. You don't want to hear this right now? OK. But remember what I said, it's gonna fuck you up. Here comes the Crippster. Later."

The Guernsey flicked its tail, turned and trotted into darkness. Steven looked over his shoulder and saw Cripps at the far end of the process line, heading his way.

Lumps of meat from the conveyor had built up in a sodden mound next to the grinder and some of them had fallen on the floor. He got up, cold and slow, and started chucking them into the machine. Waiting for the hand on his shoulder.

And it came. Cripps beside him, up against him, hard hand sliding from shoulder to neck, rubbing and squeezing.

"How do you feel, boy? Does yesterday still live within you?"

"I don't know."

Cripps laughed.

"Don't be frightened by the sickness. It lessens each time until it ceases to be felt."

"It doesn't frighten me. I just want to know what it means."

"If meaning is what you need, you'll have ample

opportunity to search for it. I'm moving you to a slaughter station."

Cripps shoved him off the stool and they moved across the floor, Steven flotsam in the bow-wave of Cripps' will.

This thing with Cripps and death trauma was impossible to evaluate. Cripps said one thing and the cow said another – and his body, when it fugued out and got sick, seemed to agree with the cow. His head, though, was greedy for change and, not knowing the correct path to take, but unable to pass up a chance at happiness, slipped into neutral and waited for the decision to be made for it.

Cripps led him on to a slaughter platform and pressed the butt of a boltgun into his hand. The slaughtermen were peripheral, the world was a grabber and a cow being manoeuvred into it. Around him there was nothing else... except the dead feeling that everything now was inevitable and beyond his control. It was going to happen – wholesale slaughter for hours on end. Not yesterday's single cow, not the separated viewing of cow death pornography, but participation in what Cripps said made these men what they were.

"You remember the feel of the gun. Good. Hold it firmly – this and your cock will raise you from your weakness. Do it, boy. I shall watch for a while."

Steven blew a hole in the cow's head, felt the animal's collapse in his own body and a fine spray of blood on his face.

Gun swings back on its chain and slaughtermen drag the still shivering cow out of the grabber and hook it up to the conveyor. Then back in again, press hard against the next cow's head and pull the trigger.

He puked over the third cow before he killed it.

Dimly, at his side, he was aware of Cripps wanking.

Aware, too, that it was over him and not the dying cows. But it meant little. He was inside himself, watching himself kill and unable to stop. Working faster and faster in sprays and fountains and gouts of blood and brain and slivers of skull and arcing jets of shit. Working fast to burn through the fever, to have it finished. But it wouldn't end and Cripps spurted come against the side of his leg, and his back and arms ached with the weight of the boltgun and his clothes stuck to him with blood and sweat and his hair was plastered down flat.

· The cows kept coming, and each one took something from him; shavings of sensitivity, perception, care. He was being robbed, violated. One of the few parts of himself he wanted to keep was being cauterised into hard scar tissue. Between waves of nausea and desperate silent pleas that the loss not be permanent, the idea crawled in that the cow in the vent had been right. He was scared. But the straightjacket of events tied him to the platform and kept his hand on the gun.

He began to phase out of perception. He dipped into long troughs of redness where there was nothing but the swaying of his body out over the guard-rail and the distant jerk at the end of his arm. During these periods he did not see or hear or taste. He knew only motion and he let it rock him to sleep, into a void where the horror of bovine death became a buffer against itself.

And then he would be back again in the immediacy of it all, feeling every ridge of the gun, seeing individually each hair on the back of the cow's head, each minute globe of blood as it danced in the air. Then, colours were concentrated, as though the dye of every object was collapsing in on itself, turning dense and hard.

On the last of these awakenings he found himself pressed against the side of a cow, down on the slaughter floor with six other men and Gummy. His dick was in it, through a hole in its hide. It was wet in there and the organs slid around unpredictably. A slaughterman held arms with him.

Gummy was shrieking down at the arse. His face dripped shit and he twitched through some kind of jig as his leathery cock splattered come over the flanks of the animal.

"Now ya know what a cow's for, dont'cha, ya little bastard? Now ya know what old Gummy meant. Thought I was just a fuck with a chewed up mouth, didn't ya?" Gummy threw his head back and shouted at the roof. "God Jesus Christ I love cows."

No one listened to him.

Cripps was alone, buggering a heifer, watching the slaughtermen through eyes glazed with the exultation of whatever truths he saw opening before him in their sadism.

The men started to make loud mooing noises, shaking their heads and bellowing deep in their chests, bringing their lips into tight O's. Steven did the same and they all moved faster and the cow's guts began to slosh.

When he came, spurting into the soggy viscera of the cow, he wanted to scream. He wanted to scream in a white hot burst words that would burn away this sin he had so greedily allowed himself to participate in. But his lungs were childhood-nightmare-paralysed as the monster races in from the hole in the wall and heads slavering for the bed and you want to yell for Dad but your body just won't do what you tell it to and you're gonna die if you don't make some sort of sound so you arch yourself until only the back of your head and your heels still touch the mattress... but it doesn't do any good.

So Steven flipped back onto the floor and blacked out.

Chapter Seventeen

It was dark. Consciousness crept back in tattered grey rags, a piece at a time, worn thin during its absence. His eyes were closed. He felt the weight of his back on the cold concrete floor, felt the weight of a black waiting silence pressing him into it. Time passed, large bodies shifted and made the air around him move, deep voices muttered vaguely. He opened his eyes, blinked, pushed himself up on an elbow. The muttering grew louder and shadows closed in. A soft hoof prodded his hip.

"Told you it'd fuck you up."

The cow from the vent.

Steven stood up in a circle of cows, light-headed and dizzy, while a single set of hooves clopped away to the edge of the slaughter room and made the lights come on. Cow faces pushed at him, a dozen, brown and pied and black. Trying to see into him like there was something they needed to know.

He squinted in the sudden brightness. The rest of the slaughter room was empty. It was late night, the men had gone. The brittle halogen light filled the room with memories of killing. He felt ill.

"You have a good time this afternoon? Do what Mister Cripps wanted you to?"

He bent at the waist and vomited.

"Oh dear. Thought it was going to make you a big strong man like all the other guys. Don't look like it right now. Tell me, man, did you enjoy killing us?"

Steven didn't answer.

"You're lucky we got the charity to dig up reasons for what you did. We could take your life away, motherfucker."

The cow rocked sideways, breathing heavily through its nose, but Steven did not feel threatened. There was more to this gathering than retribution.

"Come on, man, climb up, we're going for a ride."

"Where to?"

"Just get on."

What choice was there in the middle of this posse? Steven swung himself weakly over the Guernsey's wide back and lay flat, close to its neck, as if the life in this animal could warm away the deaths of the others.

They clattered past the empty holding pen to a vent with a grille that hung open on a single screw. Each cow got down and slid through the hole on its stomach, grunting and cursing, heaving its bulk into the space beyond. The lights in the slaughter room went out and the last cow pulled the grille back into place.

The group moved fast along the duct. Shiny sheet steel bounced their reflections back at them in ripples, golden from the low-watt maintenance bulbs that poked into the gloom every ten yards. Steven clung to the Guernsey, the breeze of their passage blowing his hair. The cows moved with a loping synchronisation, gathering momentum, merging to a single kinetic mass. There was joy in their motion, revelry in speed, grace for big bodies clumsy at rest.

Two hundred yards on, the group turned through a rent in the steel cladding and plunged like a rollercoaster down a crudely gouged tunnel, into a labyrinth of passages and chambers. Hooves rang loud on stone floors and the cows ripped out long trumpeting bellows.

Despite the still clinging horror of the slaughter room Steven was awed.

"What is this place?"

He had to shout close to the Guernsey's ear so his words weren't drowned in the clamour.

"Old sewers, old subway lines, holes in the ground, tunnels. We found them and joined them up. They go everywhere, man. Citywide. And we live at the centre. The arsehole of the city."

"This is insane."

"That we live under your feet? Why? Cripps left the first of us in the holding pen one night and we got out. Found the vent and fucked off fast. And we grew, man. Cows like pussy same as the next guy. Plenty of food down here, too. It ain't clover but, fuck, it ain't so bad."

"Didn't he come looking for you?"

"Cripps? That was in the early days when he didn't think he was a god yet. He was pissed off, sure as shit, but he didn't try to find us. Made damn sure he didn't leave anything in the holding pen again, though. And he won't stay in the slaughter room alone now either."

They jogged along the platform of an old underground station and some of the cows made train noises and chuckled at each other, nipping ears and tails and pretending it wasn't them.

"Why didn't you get out to the country?"

"Shit, man, people see us wandering around the

countryside, they'd just round us up again. And after we'd been down here a while we didn't want to be anywhere else, anyhow."

"Is it safe?"

"Yeah, but it's more than that. 'Cause our eyes are sorta on the side of our heads, running through tunnels gives us this really intense feeling of speed. Makes us feel like horses or... well, not like cows anymore.

The cows rocketed through more tunnel.

"Check out these lights up ahead. If you go fast enough it works like a strobe. See? Wild, huh?"

A string of small bulbs set into the side of the tunnel flashed by, dazzling Steven. Then they were in darkness. Total. He felt the floor sloping down, the increased speed and potential impact-mass of the cows as they lengthened their stride, felt the approach of some centre, some home, heard the animals shout.

Sudden light. And space. An explosion into openness. A columned chamber so vast that the walls were beyond the soft orange light that filtered through ancient air ducts high in the vaulted ceiling. The posse ploughed into it then slowed like their power had been turned off. Slowed and drifted with the last of their momentum into a herd that ranged out from a narrow stream in the centre of the cavern.

The Guernsey, though, had stopped near the entrance and Steven looked down on two hundred cows chewing cud, sleeping, talking together, drinking from the stream, farting, fucking, playing.

"It ain't much, but we call it home. Get down, man."

Steven slid to the dirt floor and breathed in the smells of the herd – warmth and dung and sweat, cow breath, cow presence.

"I like it here, it's like the outside doesn't exist."

"Yeah, well don't start making plans, man. This is

cow land and you can't stay."

"Why bring me here, then?"

The Guernsey walked in a circle around Steven, round and round, like a thinker pacing.

"Cripps.... See, man, you gotta understand about him. He's like the figurehead of it all for us. All the death and torture and rape are all him because he does it and enjoys it and teaches it to other men. When the first of us escaped we lived for revenge. We worked hard to build the herd, to find this place, to get into a new way of life. But all the time we knew what he was doing to our brothers on the surface. And it was emasculating knowing we couldn't do anything about it. You know what I'm saying? As long as he lived, Cripps had our balls."

"Killing him won't stop cows dying."

"Fuck, I know that. But it'll stop him living. You don't know what it's like to be in the pen watching him do that stuff, knowing your turn is coming. How it is to shit yourself with fear, to be broken even before he puts his hand on you. Take my word for it, anyone of us would die to get that fucker."

"There are enough of you...."

"Shit, Cripps is too careful. He doesn't give us the chance."

The Guernsey stopped circling.

"That's why we brought you here."

"You want me to kill him?"

"No, we want you to make it so we can kill him. Bring him to the slaughter room at night. Get him alone. Just set him up for us, man, that's all."

"It's the same thing."

"The guy's a fucking butcher. You've seen what happens because of him. Is it right? Come on, man, tell me. Is that sort of shit right?"

"Of course not. But I can't do it."

"Cripps ain't going to leave you alone, you know. You think today was bad, but he'll take you into that room again and again and you won't fucking believe what he does to you. Did you like it today? Do you want to feel that way everyday? It'll happen, dude. And sooner or later, if there's any part of you left to think, you'll want him dead just like us."

"I can't do it."

Steven shook his head, his vision blurred. He was back in the slaughter room under fountains of blood. Dicks stuck into him on every side and he was foundering, sinking fast in a bath of cow guts. The air was red and it was hard to breathe. His eyes rolled shut and he fell through the red air and, like streamers of come in hot water, the Guernsey's words stuck to him and trailed behind.

"Think about it, man. One day you'll want it as bad as us.... If you last that long."

He woke outside a storm drain at the edge of the meat plant. It was still night and his clothes were damp. He walked home. It was OK because it was too late for people.

The kitchen light was on and the Hagbeast sat at the table, fork in her fist and an empty plate in front of her. Through the window the city dawn sky looked sick – a febrile, unlaundered sheet smeared with the sweaty excretions of the dark hours.

"Where's my fucking dinner. I've been waiting all night, you animal. Where were you?"

She looked ill. The rolls of fat under her chin were grey and her eyes watered. It seemed an effort for her to remain upright in the chair. Steven was too tired to speak. Unutterably tired. He collected her plate and another and

---

stumped to the bathroom. Under the deadness and exhaustion and self-loathing there was a dim remembrance of some plan, already in motion, that must be fed and fuelled.

The bathroom was stark and dirty under early AM fluorescent light. Steven squatted over the plates. His shit came out pale and soft, in long thin strips without body. It left his arse filthy but he didn't bother to wipe, just trudged out into the kitchen again and sat down in front of the Beast. He ate without looking at her, shuddering as the rotting paste went down. But it wasn't as bad as before, tiredness and familiarity had dulled his stomach's rebellion.

The Hagbeast ate as well but had built no immunity, the first forkful made her vomit. But she didn't stop and Steven liked the wet gravelly choking noises she made as she forced herself through the plate.

"And you didn't leave me any fucking breakfast, either."

Steven finished, left the Hagbeast in a pool of puke, made it to his room and collapsed on the bed. Dog dragged over, sniffed his blood-caked clothes, then cuddled in and went to sleep.

Chapter Eighteen

Late afternoon, too late for work. Steven opened his eyes and lay wondering how he felt. He had dreaded waking, thinking it would bring with it the final, crushing ramifications of his time in the slaughter room – an inescapable knowledge of debasement. He had expected to rise tainted with the guilt of having taken life. But it wasn't so. He felt relaxed, flushed of the dross that usually chained him to indecision and fear. Like the time on the bus, he was freed of something. He felt unaccountably good.

On the way out of the flat he passed the Hagbeast, still slumped over the kitchen table. She appeared not to have moved since dawn. His breath caught and his head swam in a rush of blood. He moved carefully towards her. Could it have happened so quickly, after only two meals? He reached out a slow hand and searched for a pulse in the fat neck. Thoughts of the future jittered his arm. But when his fingers touched her skin she twitched and snorted and turned to look at him, eyes bleared and straining to focus.

"Where are you going?"

"Work."

"Cunt scum. How could you leave your Mama here like this all night? How could you, when all Mama ever wanted was the best for you?"

"You don't look good."

"Ha! Don't fool yourself, Steven. Mama knows all about the food."

She paused to suck a mouthful of snot out of her nose and spit it on the floor.

"I can take it longer than you. Where are you going?"

Steven left the flat. Her mad shriek tore at the wood of the door as he closed it behind him.

"What about my fucking breakfast?"

Upstairs, fourth-floor madness reigned still. The flat was a tip and Steven found Lucy trying to look up her cunt with a mirror. She was glad to see him and came into his arms with relief.

They sat next to each other on the couch and played at being in love. Each of them knew it wasn't real, but both of them needed the deception.

"Shall I come and live with you?"

"Soon."

Steven carried her into the bedroom because he knew it was how men acted with women. He spoke memorised sentences to her and they fucked. In the early evening they made small plans for their life together – arrangements of furniture, the colour of paint.

And they fucked again. He pumped seed into her until her thighs were slippery with it. A child was part of the happiness the TV had shown him, and by the time it was born the Hagbeast would be dead and they could all be together in the safety of his flat. His flat. HIS flat. Yes, it would be. He would make it happen. He would fill his flat with Lucy and a child and a studiously copied way of living.

In the middle of the night he got up and ate some raw meat from Lucy's fridge to make sure his shit was potent.

Lucy kissed him at the door as he left next morning. He thought of the word 'wife' and he smelt pine needles and the split cedar planking of a cabin and the brand new leather upholstery of a Jeep Ltd standing in a patch of sun. Tumblers were clicking into place, gates opening and closing along a maze, marking out a path that was his to take if only he could stay strong long enough.

He fed Dog then went looking for the Hagbeast. She was on the floor outside her room, soaked with piss and vomit and only half conscious.

"Wake up, Mama. Breakfast time."

She didn't move when he kicked her, so he took hold of one of her ankles and dragged her along the passage to the kitchen. Her dress rode up over her thighs, then further to her hips and Steven watched the scummed over, grey-haired cunt spread stickily open. Lumps of fat around her arse snagged on splinters of wood and ripped. The Hagbeast woke groggily.

"Let go of me, you shit. Let the fuck go of me."

Dried puke flaked from her chin. She struggled to sit up but Steven kept pulling.

"Not far to go now, Mama."

He heaved her into a chair and left her there, snorting back to life, while he went to the bathroom.

His shit was the colour of almond skin and almost liquid. It squirted out of his hole in a juddering stream and slopped over his thumbs as he held the plates under his arse. The meat at Lucy's had worked well.

"Here you are, Mama."

Steven served the shit with slices of white bread.

"Eat it while it's warm."

The Hagbeast lifted her drooping head and clumsily dipped a piece of bread into the steaming mess.

"You think you'll win, but there's too far to go. You'll weaken. I know you, Steven, you haven't got it in you to kill me."

"I'm not trying to kill you, Mama. I just want you to eat properly, not all that junk you used to make."

"Cocksucker."

The Hagbeast swallowed her shit-soaked bread and started to gag. Steven ate as well but, to his surprise, found it almost bearable. He was suddenly hungry and started to eat faster, sucking the shit out of his bread before he swallowed, dipping in again before his mouth was empty.

"I can still hurt you, Steven. Do you want me to show you?"

"Just eat."

"What makes you feel so safe? What makes you think you're better than me?"

"Nothing."

Steven kept his face hard but something cold took hold of his balls. Had she heard through the ceiling? Did she know about Lucy?

The Hagbeast vomited past a mouthful of bread and shit. Some of it came out of her nose. She hawked and spat then pushed her dripping face at Steven. He trembled.

"Oh, yes, there's something all right, you cunt. I can smell it on you. I'll find it, you know I will. And when I do I'll take it away from you and ram it up your arse."

"There isn't anything, I promise, Mama."

The Hagbeast was eating again, slowly and with concentration, taking small mouthfuls and keeping them down. Sweat made tracks in the dirt on her forehead and under the filth her skin was white and waxy. She was

Matthew Stokoe

having trouble holding her spoon.

"You need showing, Steven. It's been too long, you need showing you can still be hurt...."

Her words slurred then stopped. She fell sideways off the chair and lay convulsing on the floor, bubbling white foamy bile into a pool around her head.

Dog pulled itself painfully into the room, gave the Beast wide berth, and came to Steven. Steven stroked the animal's head and looked into the soft trusting eyes. He would take Dog with him into his new life and Dog would walk again and all its crippled love would be rewarded.

But right now fresh terrors gripped Steven. The Hagbeast was suspicious. Given time, a very short time, she would snout out Lucy and destroy her.

On the floor she stirred and started to get up. Steven kissed Dog quickly and left for work before she could dig into him for more clues.

Chapter Nineteen

He wanted to walk to the plant that morning but mornings, like all other times in the city except late, late night meant too many people and the agonising reiteration of how little he was like them. The bus was more controlled, people stayed separate and did not force themselves into him, they were fenced off behind the hard backs of the seats. He sat in the darkest part and thought.

They would both be waiting for him, Cripps and the cow. One wanting to continue his education in self-exploitation, the other in retribution. But he wished for neither. The thought of another bout of butchery made his stomach turn, and the importuning of the cows had no meaning for him. Today would be a day of conflicting stresses, of wills brought to bear to tug him in one direction or another. They would split his strength between them when he most needed it in tact.

Assigned to the grinder again he ruined beef all day. Once or twice he saw Cripps at the far end of the hall, entering or leaving the slaughter room, but the foreman did not approach him. Near the end of the shift, when Steven

Matthew Stokoe

was beginning to think he would escape the day without the attentions of man or cow, the Guernsey hissed through the grille and brought back reality.

"Hey, man, you look better today. You think about what we said?"

Steven stayed on his stool, but turned to face the vent.

"Not really."

"What do you mean 'not really'?"

"It's got nothing to do with me. Figure out some way to do it yourself."

"Listen, man, it's got plenty to do with you. You think he's going to leave you alone now that you've killed a few cows? You're fucked in the head. The dude's going to keep at you until you turn into one of his slaughter boys. You're going to have to do it over and over. You think you can stand that? Look how bad you were after one day."

"I recovered."

The horn sounded and the line shut down. All the men left their stations and headed for home, but Steven stayed where he was.

The Guernsey laughed.

"Big deal, you made it through yesterday. You should see what he's got lined up for you tonight. Killing cows was just the start. It don't stop there, you know."

"What's going to happen tonight?"

"Next step towards turning you into superman."

"What?"

"Wait and see. Do yourself a favour, help us get rid of him. You won't like tonight. Whoa! Time to go. Later, dude. Think hard."

The cow turned quickly and disappeared down the duct. Steven heard steps and Cripps was there, at his shoulder, smiling and waiting.

They walked silently through the deserted process hall to the slaughter room.

It was empty, the slaughter men had gone with the other workers today. Steven's feet squelched in congealing blood and made wet echoes against the walls. Cripps led him to a slaughter platform and they stood leaning on the rail looking down into the open well of a grabber at something covered with a tarpaulin.

"Well, boy, the other night was a little strong for you, wasn't it? Don't worry, I've seen it happen that way before. You can work through it. Believe me, what you think of now as horror will become glory. You will count this early sickness as small payment for the freedom it brings."

Cripps held Steven's face in his hard hands and looked into him. Steven felt like a woman, like a woman on TV melting to the demands of her lover. He did not love Cripps, though, indeed felt not the slightest affection for him. Cripps was a force that transcended personality, something to which the ordinary labels of like or dislike did not apply. The cows would call him evil, but that was a shallow description. They judged him against themselves and other men and because of this their comparisons were flawed from the start. The concept of morality had no meaning for Cripps.

No, Steven did not like Cripps. He was frightened of him, revolted by his pursuits. But here, under his eyes and his hands, the force of his will was unmistakable. At this second, despite the feeling of violation his previous killings had brought, it was impossible for him not to want what Cripps said to be true.

Cripps led Steven down to the bundle in the grabber.

"Your next step, boy. A hard one perhaps, but

necessary."

He reached over and snapped the tarpaulin away like a magician. Gummy looked up at them, his savaged lip an unpleasant colour in the bright light. Naked on elbows and knees in a pool of piss, bound with rope like a turkey. The bones in his scrawny back made sharp ridges under his pale old-man skin.

"Ya bastards! Ya shouldn't be doing this to me. I showed you how to use a cow, ya little bastard, I told you what they're for and now you're doing this to me. It ain't fair on old Gummy. It just ain't."

Cripps ignored Gummy's blatting.

"There, boy, another chance to find yourself, to realise your potential. Human this time, or almost. More potent, more efficacious. Take hold of these and force yourself through it."

Cripps handed Steven a pair of secateurs and moved behind him, close, pressing against his back, gripping Steven's wrist firmly and guiding the shears towards Gummy's arse.

"Open him up, boy. This old sack of shit will be your passport to a new world. A world of men that does not play home to fear."

Gummy tried to look over his shoulder at them.

"You're a mad bastard, Cripps. You were mad the day ya got here and you've gotten worse. I'm not a fucking passport, I'm an old man. I'm just an old man...."

Gummy started to snivel and repeat himself.

"Don't wait, boy. Don't lose your nerve. Trust me and open him."

Cripps pushed Steven's hand so that one blade of the secateurs slid into Gummy's anus. Gummy tensed and whimpered and pleaded with them to stop. But things had gone too far for Steven to listen. Cripps breathed into his ear.

"Now, boy. Don't wait any longer."

He squeezed Steven's hand closed. The secateur blades made a soft crunching noise as they scissored together and cut through the muscle of the rectum. Gummy screamed and vomited and sprayed blood from his ripped arse. The ropes bit into his forearms and thighs.

Steven closed the shears again. And again and again, up from the arse and along the right side of the spine. Gummy's lower back opened to a rear view of guts. Easily. Then Steven hit the ribs and the going got tough. He had to apply a lot of pressure and twist them sharply sideways to make them snap. Spurted blood dripped from his face, off the point of his nose and the end of his chin, and back into Gummy.

Gummy lost consciousness and stopped screaming. Steven clipped him open all the way to the base of his skull.

When it was finished Steven stood looking at his handiwork, knowing what he had done, panting and unbelieving. Nausea came and he vomited into the gaping body. With the puke went his strength and he fell backwards on to Cripps who lowered him carefully so that he sat with his back against one end of the grabber.

"Easy, boy, easy. Sit there and rest. The first man is the greatest hurdle and you are past it. Sit there and feel the glory of it fill your body. You have done what only a very few ever dare try. You have changed yourself. This and the time with the cows has changed what you are. You'll see. When the sickness passes you'll see."

Steven wasn't listening. The enormity of what he had done overloaded his senses and all he could see or hear was the frozen black void in which he hid. Outside there was pressure against his skin as Cripps kissed his cheek. Then he was alone, with the darkness and the silence.

Matthew Stokoe

Cripps was gone and time passed. And when enough of it had passed there was standing and walking, on to a bus and off again.

Until the fourth floor and Lucy's flat and a long slow dissolve into function.

He found himself by her bed, looking down at her as she slept. On the floor, surgical photos and texts had the glossy red look of pornography. At this moment, though, they meant nothing – not madness or despair or titillation. He was conscious only of the desire to be next to her, curled against her. Asleep.

Chapter Twenty

Sun through an open window woke him. In its light he felt bright and clean and new. Lucy was warm beside him and he felt no pain – no small waking aches, no apprehension at the day to come or fear of the years massed behind it, no anxiety at decisions to be made. He stretched and flexed the muscles of his body, they were hard and wanted action. On this morning the road to his future was clear and sharply marked. The killing of Gummy still existed but its horror had become part of him. It was not as it had been yesterday. While he slept it had changed, been absorbed, so that it was now a dynamic heart that beat deep and sure and untroubling within him. He had expected it to suck the life out of him like a tapeworm, instead he felt reinforced, strengthened, capable.

He looked at Lucy, still sleeping, sunlight on her hair, and realised that murdering the Hagbeast by degrees was an unnecessary prevarication. In the clear, stripping light that filled the room he saw he had been weak and frightened, but he saw also that these emotions were now outmoded. And as hard as he looked he could see no barrier to action.

He was eager to start.

Naked out of bed he felt like a god, reaching high into the air, a theomorphic diver stepping willingly to the edge of a cliff for the plunge into transforming waters. His movements were sure and exact, he marvelled at them as he dressed.

Lucy slept on.

Walking downstairs – each step certain and excited. He knew what he was going to do. This morning would corrode the past, leave it a honeycombed and fragile shadow falling to pieces miles behind him. Why had it taken him so long to act? The resolution of everything he wished was so obvious and easy. He didn't understand.

Then he was through the door of the flat and understanding didn't matter.

He wanted a knife but the kitchen was dark. The Hagbeast had covered the windows. Steven stood in the doorway groping for the light, wondering what the bitch was up to. He could sense something in there, something heavy and waiting. He heard movement at the far end of the room, the start of motion, a mass shunting forward, gathering speed too quickly. Like a train, or a bull, or a rhino.

Switch clicks down. Light. And there she is, three quarters of the way across the room already. Too fast and too close. Thundering. Arms pumping, mouth sucking in air and spraying spit, blasting that body, that unstoppable bulk straight at him. Time just to think SHIT! and try to make it back out into the hall. But not enough time to do it.

The Beast hit him hard in the back, bodyslammed him face first into the floor of the hall. Sprawled on top of him, pinning him there, grinding his face into the wood. Impossible weight crushing him. Sagging fat of belly and breasts engulfing and smothering, preventing the use of arms or legs. Far up the hall Dog looked fearfully out of Steven's room.

"Too late, Steven. Mama beat you to it, you little fuck." .

The Hagbeast lay on top of him, shouting at the back of his head. She had a length of coarse rope and she looped it around his neck.

"Did you think I'd let you do it to me, you moron? That I'd keep eating your shit until it killed me? Gutless cunt. Wouldn't dare do it properly, would you?' Well, Mama said she could still hurt you, and now she's going to show you how."

Her stink was overpowering – shit and stale sweat and rotting cunt blood. Steven thrashed and humped his body in short jerks off the floor, but her fat absorbed his movements and he could not escape. She lay immovable, slowly tightening the rope, pissing over the backs of his thighs in her excitement.

He felt his throat closing, pressure building in his head, making his eyes bulge. The Hagbeast had her face nuzzled close against him and he could hear her grunting with the rope. Down the hall Dog dragged itself towards him, stumping one foreleg after the other as fast as it could, panting, twisting its face with the strain on its heart, eyes locked on Steven, begging him not to die.

"See, Steven? See how Mama is still so much stronger than you? Shit was too slow, dummy. You should have known. Can't sneak things in the back way, not with Mama."

She hauled tighter on the rope. Steven's face went dark and the veins above the rope got thick with dammed blood. His vision was starting to fade but he could see that Dog was close now.

Matthew Stokoe

Yes, Dog was almost there now. It was going to save its master, the source of all love, even if it meant death. Even if the race along the hall burst its little heart and made blood run out past its dog glory of sharp white teeth.

And the Hagbeast didn't know. With her head pressed so close to Steven she could not see it coming.

Steven felt himself draining into the floor, going cold and slow and heavy. His lungs sucked against nothing.

The Beast was laughing.

Dog in a mist, but close. Steven could pick out the white whiskers in its muzzle and the darker hairs further back and the foamy dabs of spit along its lips. Dog's head expanded as it came, filling his vision until there was nothing in the world but this dog face and the love pouring from its brown eyes. And the Hagbeast's laughter in another corridor somewhere very far away.... And the absence of air.

Dog was past Steven's face, climbing awkwardly on to his shoulder, pushing its head forward, up up up to the Hagbeast. Still she didn't see. So Dog got into position and with the last of its strength opened its mouth and closed its eyes and sank its sharp white teeth into the Hagbeast's neck and held on as she shrieked and reared up. And let go of the rope.

Steven rolled out from under her and gulped air while she twisted and battered at Dog, trying to dislodge it from her throat. He wanted to move, he wanted to save his dog and tear his mother into bleeding pieces of flesh, but his air-starved muscles would not respond. So he knelt, slumped against a wall, dry retching, breathing in sobs, and watched the Hagbeast rip Dog from her neck in slow motion.

She held it like a spear and pulled her arm back over her shoulder, then paused and turned to smile at Steven. He tried to scream but he couldn't and the fat arm shot forward and drove Dog's head into the wall. Steven followed the whole of the movement, from the start of the arc to the explosion of blood and brain against the flaking plaster. Dog held his eyes through all this terminal journey and they were loving and sad at the same time. It looked like Dog was smiling, just a little bit. Until its eyeballs burst.

Then muscle power came up in a rush and Steven was a blur of vengeance across the few feet that separated him from his mother. His body flowed like water, unfettered by thought, channelling a lifetime of hate. He struck the grinning Beast mouth with his elbow and she collapsed face down in the stuff that leaked from poor Dog's shattered head. Steven was back on track again, back on the path that had started this morning at Lucy's, the sure straight path that led from lonely TV nights, through the slaughter room and Lucy, to here and on to dreamland.

He lifted the Hagbeast under the arms and dragged her into the kitchen.

She woke sitting on her knees on the floor, bound and immobile. A rope ran around her forehead to her ankles, pulling her head back, stretching her throat out straight. The marks of Dog's attack were turning blue. She had difficulty speaking, but that didn't stop her.

"Do you think Doggie enjoyed that? I did. Fucking mongrel. No more dog suck-offs for Mama's best boy now, eh?"

She cackled and tried to twist her head to see Steven better, but the rope would not allow it and her eyes rolled in jagged circles.

"What's the rope for, Steven? You know Mama's going to be mad if you keep her here too long. Better let her up right now. Ooo, Steven, what's that mark on your neck? Looks like a rope burn. Let Mama have a look at her

Matthew Stokoe

poor boy."

The Hagbeast stopped abruptly and ripped out a stream of phlegmy laughter. She started to choke and spat in a high curve back over her head. When she was breathing normally again Steven wedged a small block of wood between her right rear teeth, jamming her mouth open wide. She gurgled and looked frightened.

"You were right, Mama, shit was too slow. But I don't think you're going to like the alternative."

The pliers were heavy and had rubber grips. They were a solid man's tool and he felt very confident holding them. The Hagbeast still had most of her teeth, they were a little yellow, but they were there. He started with the small lower incisors at the front.

The knurling on the nose of the pliers rasped off small pieces of enamel even before he applied much pressure. The Hagbeast whimpered and tried to swallow. Steven closed the pliers firmly and jerked them forward, splintering the tooth and snapping it off just above the gum. Her body went hard with the pain and she screamed. Blood ran backwards over her tongue.

Steven let her relax a little before he crunched the pliers closed again.

The teeth at the sides were harder to break and some of them came out by the roots. There was a lot of blood and Steven had to push her over on to her side twice so she wouldn't choke.

He was sweating by the time he finished. Fragments of tooth were embedded in the soles of his shoes, they grated against the floor when he moved. The Hagbeast was still conscious but her eyes were glazed and she had stopped making noises. Her gums were a pulpy red mess with the sharp remains of teeth poking through. The front of her dress was soaked.

Steven threw the pliers in the sink and picked up a file to smooth down the spikes. The Hagbeast passed out at the first screech of steel across enamel. It made it easier for Steven to finish his work.

When she came to he stripped off his trousers and pants and looked down at her for one last moment – this mother who had never been a mother. She bubbled thickly up at him but he couldn't work out what she was trying to say.

He stuffed her nose with wadded toilet paper, then backed up to her until her wedged-open mouth was pressed between the cheeks of his arse, tight around his hole. He used a roll of industrial adhesive tape to bind her there, wrapping it round and round, over his abdomen and behind her head. The seal was air tight and he could feel her shake as she fought for breath she was never going to get.

The shit was packed in his guts – twenty five years of terror and loneliness, of brutality and an endless rain of hate. He breathed in deeply, tightened the muscles of his stomach, and shot every ounce of it in a thick pole down her throat. The Hagbeast thumped up and down, vibrating in a mad death dance as the shit blocked her from mouth to belly. Steven had to reach round and hold on to her head until she went limp.

He dragged his clothes back on, sat at the breakfast table and looked at her. It was done. The obstacle was removed. He would bring Lucy down and there would be a home here for him at last. And if Lucy and he could not be like others they would at least approximate the happiness others had. Lucy would watch TV with him and learn how to live. They would scale down and copy what they saw, and they would call it contentment.

Although he was staring straight at her, each minute that passed made the Hagbeast seem less real. It was as though she were fading back into time – almost as if killing had expunged the memory of her mistreatment.... But no, she was here now and she had been there through all those years. She had made him what he was, he would not forget that.

And he would not forget the pride that coursed through him as he sat there. He had done what he thought he could never do – he had destroyed the source of his misery. And he had done it powerfully and like a man.

He left the Hagbeast where she was and went to get Lucy. They spent the rest of the day moving her things into the flat.

Chapter Twenty-One

Evening. They were drinking coffee in the kitchen. Lucy had her eyes on the body.

"Will you let me have her, Steven? Something that ugly must have stones in her. Can I look inside? It can be a present to mark the start of our life together. Let me open her up."

Steven sighed. This weirdness of Lucy's unsettled him, it did not fit with his picture of how things should be. In his dreams he'd seen an instant normalising of behaviour when they began living together. It was clear now that things were going to take a little longer.

"OK, but she goes tonight."

Lucy kissed him and unzipped her wallet of scalpels. He left her to it and headed for the bedroom, picking up poor Dog's body on the way, holding it close. He needed to sleep for a while.

When he woke at 2AM he was giggling. For a long moment he was inside the TV, running across the green fields of Dad's crops to a white sunlit house with animals playing all around where Mom was waiting to hug him to her big soft chest and say *Gosh I love you so much Johnny, I could eat you right up you're so scrumptious.*

Then he was back in the room, the room that would have to be changed so much. The TV was on and everything it showed looked possible.

Out into the hall. Into the flat. Into HIS flat now. The walls glowing with pleasure to see him how they always wanted – lord of the place, uncontested and safe. And he did feel safe. He was certain of everything. In here, with the Hagbeast gone, his dreams of love and comfort would harden into reality around Lucy and himself, undisturbed by the currents that tore at the world outside.

He knew what he would find in the kitchen and it was all right. It was part of a necessary transition.

Lucy stood crying by a pile of shredded meat that had been his mother. The mess on the table was unrecognisable, every organ and every piece of flesh had been stripped from the carcass and minced. Many of the bones had been splintered and torn from their holding cartilage, even the skull was open and scooped clean. The face hung from it in a peeled flap of skin, like an inside out halloween mask.

He held Lucy and stroked her hair, whispering reassurance, smiling gleefully to himself over the top of her head because her search through the foulness of the Hagbeast had been fruitless. Now that she had looked inside a human, picked one apart with her own fingers and found nothing, she was more his than ever. This final, unequivocal loss of hope would force her into the hidey-hole of life with him and a child.

She clung to him all the way back to the bedroom and when he fucked her she held on and didn't let go until she fell asleep.

He left her curled in damp sheets, twitching and

murmuring unhappily to herself, and lugged the sodden remains of the Hagbeast up to the roof in black plastic garbage bags.

When he climbed out into the night the city was young again as it had been during the secret visits of his youth. He stood by the low wall at the edge, drinking it in, caught in its regenerated spell. Neon, distant music, even occasional laughter, floated tantalisingly about him.

He leant against the wall and looked out over the endless sprawl of buildings. Two bricks fell away and smashed on the empty street below. He felt like a king, like he could command the buildings to tear up their roots and march away if he wished. He was beyond and above it all. Only a week ago the sight of so much living would have crushed him. What had given him this strength?

Links of conclusion formed chains as he tracked backwards through increases of power. He held his breath.

Cripps was right.

It had been killing, obviously, that had allowed him to reach this position of self-determination. He killed cows and he was able to start poisoning the Beast. He went further with Gummy and was able to kill her outright. The slaughter sessions had worked.

He went downstairs and came back with a can of gasoline and Dog's dead body. Dog had waited all its life for Steven to take payment for a broken back and it was only fitting that the remains of the animal should witness this final destruction. Steven wedged the stiff bloodstained canine between two chimney pots and made sure it had a good view.

The Hagbeast meat made sucking sounds sliding out of the bags and some of the bones tore holes in the plastic. It took all the gas in the can to set the mess alight.

Beast barbeque.

Muscle sizzled, wads of fat caught fire and burned along with the gasoline. Then it all got black and started to smoke and the pile sank in the middle and collapsed in on itself.

At the end of it all the Hagbeast was a greasy smear on concrete and powdery lines of bone ash lifting on the night breeze. Dog looked so settled that Steven left it where it was between the chimneys, staring with its burst eyes out past the Hagbeast remains to the pretty lights of the city.

Chapter Twenty-Two

The weeks that followed were happy. Lucy recovered from her disappointment at the emptiness of the Hagbeast and buried her horrors under a fevered procession of decorating, fucking and moulding herself to Steven's vision of life. She watched TV with him late into the night, taking notes and listening carefully as he pointed out particularly relevant scenes and networks of emotion. Together they painted and cleaned the place, destroying every trace of the Hagbeast and the life that had been lived there before. They made a simulacrum of all the perfect family-show houses – *Brady Bunch*, *Happy Days*, *Cosby Show* – so they could live perfectly themselves.

The flat opened out and breathed again, and the sun changed minutely its course in the sky so the rooms were filled dawn to dusk with its shine. There was cleanness and order and warmth and companionship.

Steven had made his dream real. Lucy was pregnant and eventually there would be a child, and with it would come the family he had seen night after night on TV. He would have to get another dog.

Matthew Stokoe

Although he was still prone to the great wretched comparison of lives – his and the rest of the world – he felt at times superior to other men. They were blessed with happiness from birth, but he had had to force his into being with the strength of his own hands and will. And when he padded through the flat in the early morning, savouring the completeness of his satisfaction, he knew he had crafted well enough to be worthy of TV.

But things did not stay that way. As time passed he became less sure of himself.

It started three weeks after the Hagbeast's death – a nagging anxiety that daily became more definite. At first he dismissed it as a reaction to sudden change, but the unease grew until each morning was a dreaded thing, bringing as it did a fresh increment of fear. The confidence of the first weeks left him and an impotent knowledge of the thousand massing things that could destroy his new life took its place.

His will alone maintained the world within the flat and the strain of resisting its collapse became unbearable. So many things could happen – Lucy might crack irreparably, the building might fall, he might wake one day to find he simply could not support his new freedom. And money.... The rebirth of the flat required funds and he had not been to the plant since that night with the pliers. It was too much for a weak man.

But he had been strong before. He had had the strength to kill his mother.

It took a week of snivelling through early morning hours until he understood what he needed. A death. Killing. Killing had given him the strength to start things and he needed more of it to continue. He needed another blood burning injection of certainty. He needed what Cripps had shown him.

Chapter Twenty-Three

Steven left very early for the plant, before the streets filled. Left while the sky was still an orange-scummed blackness so that the bus would be empty. He felt flayed, as though every paranoid receptor he possessed was trained on the gulf between himself and others. His brief escape from inferiority only intensified the pain of his return.

He made it, but it was hard. He kept his eyes closed and pushed himself tight into a corner of the dirty vinyl bench at the back of the bus. He counted stops until it was time to dash into the coolness of the dawning city. Trash cans in an alley at the side of the plant hid him until the gates opened, and then he made another dash.

Inside, back at the grinder, it was better. The process hall, with all its ghastly content, held some degree of familiarity that made the world easier to bear. A month's absence seemed to matter to no-one. He clocked on as usual and was assigned as usual, sat on his stool and humped meat as usual. Once, far down the hall, Cripps stuck his head out of the slaughter room, looked straight at him and smiled, nodded, then disappeared again.

At lunchtime the Guernsey pressed its face against the ventilation grille and spoke to Steven.

"You're back, man. We been waiting a long time and I gotta say our faith was getting shaky."

"What faith? I told you I wouldn't do it."

"Well, us cows got that old intuition, we knew you'd change your mind."

"Who says I've changed it? Maybe I'm back because I need the money."

"Sure, dude. And maybe I take it up the arse. We saw you hiding out there at the side of the plant this morning, twitching and jumping. Cold turkey, man."

"Bullshit."

"That supposed to be funny?"

"What do you mean?"

"Forget it... You can't kill without getting infected. It don't have the effect Cripps says, but it gets under your skin in other ways. We warned you."

The Guernsey's voice got harsh and Steven saw the muscles of its face tighten.

"Now stop fucking about and tell us how it's going to happen."

"If it happens at all it has to be how I want it."

"Set it up any way you want, man. Long as it happens we don't give a shit."

"I get to do it."

"Huh?"

"You can watch and be there, but I do the killing."

The Guernsey was silent a moment considering this, then:

"My, my. What happened to the boy who puked when someone else shot one of us measly cows? I don't know, man.... I wanted to do that fucker myself. The herd wanted to be involved."

"You won't get him without my help. You said yourself he's too careful."

"So we'll wait for him to get careless."

"You could wait forever and you still wouldn't get him. You know that. Make your choice – let me kill him, or let him stay alive."

"I don't like it, man. You're taking away something that's ours."

"If he dies, he dies. As long as you're there to see it happen what difference does it make? What can you do with those hooves, anyhow?"

"Give him a fucking good stomping. What do you mean?"

"Look at my hands."

Steven rippled his fingers.

"I can do things you can't. I can hurt him more."

The cow studied Steven, chewing cud and breathing noisily. It swallowed and stamped.

"OK, man. I ain't happy about it, but ok. And we gotta be there, real close, understand?"

"From start to finish."

"Good. What's the plan?"

"Watch the slaughter room at the end of shift. Come when you see us."

The Guernsey nodded and faded back into the shadows of the duct.

There was a little time left before the meat started to roll again and Steven used it to set up the evening.

Cripps was alone in the slaughter room staring at a wound in the head of a dead cow. When he saw Steven he straightened and walked quickly to him.

"Hello, boy. You look well. You look as a man should – without fear of killing, without fear of himself."

"I want to do it again."

Cripps embraced him.

"Of course you do, boy. Of course you do. You were gone a long time."

Steven spent the rest of the afternoon trying to work off a steadily building stress. He threw meat into the chute of the grinder as fast as the machine would take it. He wanted to run the length of the hall, to tear around it, screaming and shouting and smashing things. But it wouldn't have done any good. Nothing could stand in for the ecstasy of becoming another person.

Chapter Twenty-Four

When the men had gone home and the plant was quiet, Steven returned to the slaughter room. Cripps was there by himself with a cow in a grabber. Somewhere at the back of the room water dripped slowly from a tap on to concrete. The halogen above the grabber was on but the other lights were dead and it was hard to see anything beyond the purple-bright cone of light.

The cow in the grabber gave Steven an unexpected twinge of disappointment. He was disappointed it wasn't a man. It would have doubled the night's score.

"I have it ready for you, boy. Ah, how willingly you come now. How strongly you walk. This one will bring you no sickness."

"I'm past getting sick."

"I know you are, boy. Here, take a knife."

Cripps' face stretched into a grin. His fly was open and his dick poked through it, pointing at the roof. Steven held the knife, an electric flensing tool, wondering how long he would have to wait. Did the cows expect some kind of signal?

"I'd rather use the gun."

"As you wish. Give me the knife, I shall use it."

Cripps pressed the stud on the handle and the eighteen inch blade started to buzz. Steven swung a bolt gun down from the slaughter platform and charged it by pulling back on the lug behind the grip. The hiss of compressed air into the weapon was sharp in the empty room and the cow in the grabber jerked. Cripps walked round and stood close to the animal's head. He nodded to Steven.

"The jaw."

Steven looked behind him but could see no cows approaching through the surrounding blackness. So he fired the gun into the animal, into the lower jaw, midway between the chin and the cheek. The cow squealed through its nose and lunged against the grabber. It was not mortally wounded but its mouth was smashed and hung blood-slobberingly open.

"That was fine, boy. No hesitation at all."

Cripps winked at Steven then put the blade of his knife between the dribbling lips and cut backwards using both hands through the cow's head – through the corners of the mouth, back under the eyes, deep into the skull about an inch below the ears, then out through the back of the neck into air. The entire top of the animal's head came away in an eruption of blood. Cripps shrieked laughter and pushed his face into the red fountain, cupping his hands around the brain half that lay cleanly sliced in its shallow dish of bone, jerking it away from the spine and lifting it out. He found the other half somewhere on the floor and handed them both to Steven.

"Do me this service, boy."

Steven held the pieces of brain like a bun while Cripps slid his cock backwards and forwards between them

---

until he came. A thick grey paste built up in the folds of his foreskin and the semen that splashed across Steven's wrists was tinged pink.

After he finished, Cripps leant against the side of the slaughter platform, wiping his face, laughing between gasps. His hair was thick with blood, the lines of his face filled red. Steven stood beside him, toying with the boltgun.

"You make me proud, boy. You had so much further to come than the others. You prove it works for all men."

"Yeah, it works."

The cows arrived. Steven could see them dimly beyond the curtain of light, moving carefully, quietly, closer. Cripps had blood in his eyes and did not notice them immediately. Steven readied himself.

"But it works better with men."

"That it does. These beasts are poor substitutes. What was that?

A hoof scraped on concrete, then silence.

"There is something here with us. We must leave."

Cripps pushed himself away from the slaughter platform then stopped dead as the cows moved into the light. There were ten of them, in a solid half circle that curved around the grabber, trapping him.

"Use the gun. I have suspected this for some time. They will kill us if we give them the chance."

Cripps was not frightened. His movements were smooth and unhurried as he pressed the switch on the knife and raised it before him.

"It won't be them doing the killing, Cripps."

Cripps frowned over his shoulder, puzzled. Steven bent quickly forward and fired the boltgun into the side of his knee. Cripps grunted and dropped the knife. The joint was shattered, white bits of bone stuck through skin and trouser fabric.

Matthew Stokoe

"What is this, boy?"

"They want revenge. They've been waiting a long time."

"But you? What are you doing with these animals? You should be with me."

"I'm doing what you showed me. Gummy wasn't enough. And my mother, too. I killed her, but it didn't last. I need more and you're it. You should understand, Cripps."

Cripps steadied himself against the rail of the platform. His voice was quiet when he spoke.

"I taught you the secret, and yet you can do this?"

The Guernsey stepped forward and kicked Cripps' shattered knee. Cripps hit the floor.

"Shut the fuck up, man. It's time to go. Come on, Steven."

Steven picked up the flensing knife, checked that it was well charged, then dragged Cripps across the floor to an open vent. The cows slid through first, Steven pushed Cripps after them, on to the back of the Guernsey.

Then he was in the duct himself, shoving the vent grille back into place, climbing up behind Cripps as the posse started to move. Cripps tried to raise himself but Steven held him firm and clubbed him unconscious with the haft of the knife.

The cows did not speak or joke on this journey. They moved quickly, intent on their destination, and in this short time between capture and kill Steven could relax. The swing of the animal's stride, the rhythmic sideways rolling, the play of muscle under hide lulled him. He closed his eyes and let himself be carried, remembering the feeling of strength he had discovered the night the Hagbeast choked on his shit.

He wanted this thing with Cripps to be over so he could carry that same feeling back home to Lucy and the flat. He wanted to do it now, here on the Guernsey's back – but he knew the cows would not stand for it. He would have to wait. Just a short time.

When they hit the central chamber the waiting herd shouted. The place was dim like last time and there was a tension in the air that turned the warm earthy smell of manure and cow breath sour. The cavern was no longer a place for unguarded resting and loving and children's playing. The edges of things were strained and sharp, and where the herd had been so unconcernedly scattered before, they were now ranked tightly at the edge of the stream, watching Steven's approach.

The posse trotted forward to join them, but the Guernsey stayed slow, moving with measured steps as though this moment carried gravity in the cow scheme of things.

The herd parted before them, opening a corridor to the edge of the stream. As he passed along it every cow tracked Steven with its gaze. He held himself straight under the scrutiny, aware of the need for ceremony, knowing that for these gathered animals the killing would section the past from the present. While Cripps lived they were less than they should have been. Today they wanted things to change.

Steven dismounted. Cripps was muttering broken sentences to himself, clawing his way into consciousness. An active Cripps would only make things more difficult, so Steven worked fast.

There was a tangle of rope at the edge of the water, no doubt placed there by the cows for his use. And under it, four short lengths of wood and a heavy stone. He heaved Cripps from the Guernsey and dropped him on the hard earth floor, looked once at the crowding animals, then staked him out, naked and spread-eagled, like on cowboy

Matthew Stokoe

films.

By the time he had finished Cripps was awake, looking up at him with a half smile like this was just the endplay in some game and he was happy to be involved. Steven looked back at the man who had opened the door to his future, the man who had learnt the secret of a nightmare and, having learnt, pulled it close about him and fed off it, forsaking all other food until he grew strong enough to share it.

"I knew you had it in you, boy. You make me proud. You will be a man above all others after this. A man like myself."

"I'm about to kill you, for Christsake."

"And you do it for a reason, don't you? You expect something from it, tell me you do."

"Yes, I expect something from it."

"There! You are my proof. When a man is truly free he is capable of anything that serves him. I knew it, but I did not dream I would see such an exhibition of it as this will be."

The cows shifted impatiently, the Guernsey stepped forward and spoke close to Steven's ear.

"Hurry up, motherfucker. You wanted this for yourself, get on and do it or I'll take over."

Steven looked down again at Cripps.

"I'm ready, boy. Show these meat fuckers what we men are capable of. Show them the power we carry inside ourselves. Come on, boy, don't keep me waiting."

Cripps was shouting. He shouted louder when Steven fired up the knife and made it hum.

"Don't keep me waiting, boy."

Steven started on the left arm, slicing shallowly through the tricep, down from shoulder to elbow, taking away long slivers of meat until the bone showed. There was a lot of blood – it would be over too soon if it went on like this. He stopped and used what was left of the rope to tourniquet Cripps four times – once at each shoulder and once at the top of each thigh by the groin. The flow of blood from the left arm slowed.

Cripps did not scream or show pain beyond a tight squinting of the eyes. Instead, while Steven was close to him tightening the ropes, he whispered urgent endearments – encouragements that were the last rites in the faith of himself.

They meant nothing to Steven.

He started with the knife again, working from shoulder to wrist, cleaning both arms of flesh. Then on to the legs, letting bone see light for the very first time. Cripps lived through all of it, but his eyes got dull and he did not see the cuts of meat that piled in outline around him like a snow-angel.

Flesh-angel.

But Steven saw it all, each increment of death, each weighting of the scales towards that long black fall. And he sucked up each degree as it came, storing them away until there were enough of them to make up one whole death.

The slowness was only for the cows. Torture was not important to him, in fact it was annoying. If he had had the choice he would have strangled Cripps, felt the life go in one quick rush up his arms – in one slamming, changing wave.

Cripps was a torso stuck with four white bones that ended in hands and feet. But his chest still rose and fell and there was more to do. Obvious things like opening the stomach and pulling out handfuls of guts, like splitting the penis lengthways and cutting off his balls, like using the very tip of the knife to cut out his eyes. Things the cows watched in silence, breathing hard and swallowing rapidly, urging him on with a silent desire to see it done.

Somewhere in it all Cripps died and Steven became as he had been immediately after the Hagbeast's death. The destructive uncertainties of the last week ceased to exist, had never existed. He was back in charge and life looked smooth and straight and certain again. There was no question of his ability to cope with it.

At the end of the last long slice up through the chest Steven straightened, dropped the knife and looked at the wall of cows. For a moment it was like they all stopped breathing, then they knelt and bowed their heads – all but the Guernsey who stepped quickly to his side.

"OK, you did it. Get on, it's time to leave."

Steven swung himself up on to the broad back. He was puzzled by the Guernsey's haste, but there was nothing to keep him in the cavern. He held tight to the rolls of skin around the animal's neck and they cantered quickly from the chamber, unaccompanied, out into the maze of tunnels beneath the city.

The Guernsey took him to the same storm drain as before, a shattered tube of reinforced concrete that opened into a pit in a garbage-strewn empty lot.

"Get what you wanted, dude?"

"You wanted it done."

"Yeah, but you didn't do it for us. I was watching. You needed it just as bad as them back there."

"I'm going."

"OK, but they're going to want you back sometime."

"What do you mean?"

"You didn't see them? They were on their knees."

"So?"

"It fucked their heads, man, a whole lot more than I figured. This ain't the end of things, not by a long shot."

"Whatever you say."

Steven climbed out of the pit. The Guernsey was speaking from · a different world and its words meant nothing. They were for somebody else, not the self-contained god who left them behind as so much noise and strode through the twilight along city roads that glittered with the phosphorescence of recognition. He knew these streets, had pursued them through dreaming Hagbeast nights, had plotted and examined each stone and dollop of tar that supported the TV lives of all the people who trod them. Tonight they opened effortlessly for him, empty and softly glowing, leading straight to his flat.

Chapter Twenty-Five

Everything was perfect, clean and warm and bright. Lucy cooked for him and held him and they sat together through the long nights. The days were ordered and serene. He woke and they drank coffee together and breakfasted on fruit. Lucy, in a fresh towelling robe, raised up on tip-toe to kiss him each morning as he left for the plant. She gave him his lunch and waved goodbye and waited for his return. And when he came home the kitchen was wrapped in the welcoming smell of baking.

He was safe. He had succeeded. And it lasted a long time.

But safety brought about change. Initially, the drudgery of grinding meat at the plant had been lost in the excitement of exploring his creation – living as a man with Lucy was a buffer against its detrition. But the day came when Steven saw that he was better than the job. And from this day the idea of time spent at the plant became repellent.

The endless procession of lumps of beef disgusted him, the swinging carcasses on their hooks were odious pendulums that marked out wasted time. A man who could find it in himself to murder his mother and Cripps was surely meant for better things than the arse end of a production line.

When the monotony reached a point where Steven thought he could no longer contain his rage the cows, absent since Cripps, came again. The Guernsey gave instructions through the ventilator at lunchtime, and dusk in the dead plant found Steven in the duct, climbing on to the animal's back.

They were alone and the warmth of the beast was welcoming – a blanket of comforting memories. And more than that – as they raced through tunnels, Steven felt a resurgence of energy, a freshening of the excitement that had coursed through him with the killing of Cripps and which the dragging hours grinding meat had dulled. The breeze and the movement of the Guernsey under him sloughed away the scab of forgetfulness that had grown over his sense of greatness.

"Ain't this just like I said? That you'd be down here again?"

"You came for me. What do you want?"

"Not me, man, I don't need anything from you. The others do, though. Or think they do. They changed after you did Cripps. Like his death started something that'd been waiting to happen since we came down here. They're fucked up good now."

They moved through a dried-up water main and the Guernsey's hooves clattered hollowly ahead and behind. Steven took a deep breath of the musty air and stretched his arms wide until they almost touched the old brick walls.

"You sound happy about it."

"I just got a little more distance on things, that's all. I evolved faster down here than they did"

"Why do they want me back?"

"Anxiety, man. That and some sort of misguided faith in you. After Cripps there was like this massive upsurge of energy in the herd. They knew they'd changed, but they didn't know how. Everyone stayed close, they couldn't handle being alone. See, what scared them was they didn't have an identity any more. They'd spent so long hating Cripps and wanting him dead they were lost when it actually happened."

"What did they do?"

"Stampede. Only way to burn off that energy. Whole herd, kids and all, running under the city until they're too tired to move. But it don't really work 'cause when they get back they sit around and start thinking and get fucked up all over again. We should be spreading out through the city, expanding the herd, but they won't do it, they won't split up and leave each other."

The Guernsey looked quickly back over its shoulder.

"They're expecting answers, man."

Steven was silent, but the feeling of imminence that had risen within him at the start of this ride grew stronger. Grew from a soft-edged augury into a certainty.

He smiled and settled back on the cow, he felt like laughing. Crazy Cripps. Mad fuck-up blood-bathed torturer, dead by Steven's own hand, had seen the routes and the systems of life with x-ray eyes. He knew how the game was played but thought the slaughter room was the whole world and so had never directed his incandescent amorality beyond it. Despite this, he'd taught Steven that things exist to be taken by man, that a free man lives as the centre of his universe, directing all things to his ends, totally.

And if the Guernsey didn't lie, Cripps, by his death, had provided one more thing to take. He'd opened the cows to exploitation.

The chamber had deteriorated. The vital parochialisms of a community safe in its own territory were gone. No lounging, no playing, no joy in association. Piles of dung, scuffed and trodden, littered the open floor. Neglect hung heavy in the air and mixed with the stale ammonia tang of urine.

At the centre the herd was a steaming, hoarse-bellowing conglomerate, tight packed in violent motion around something torn and rotten propped against a pillar. Sweating cow heads pointed into the heart of the vortex, going round and round and round.

The tumbling, end-over-end impact of hooves rolled through the chamber, bouncing off walls, folding in on itself, returning to the centre, only to be thrown out again louder and more dense with frustration. Steven, on the Guernsey at the entrance, sat straight and breathed it in, let it wash about him and thrill him with the electricity of its abandonment.

"Told you, man. They're going to be plenty pleased to see you."

The Guernsey walked slowly towards the spinning cows and Steven went looking inside himself for streams of words, the exact right ones that would thud into empty bovine heads and lodge there, barbed and solid, so that when he hauled back on them he could be sure the cows would follow. He found a surface blankness, but he did not give up. Somewhere below consciousness the cows must know what they wanted. His voice would reveal it to them, it would give them a flag they would recognise and follow.

The herd saw him coming and slowed. Dust settled. Steven rode through them to the pillar and saw with no surprise Cripps' tattered skeleton.

Sound of cow-panting all around him, great chests heaving. Ripples of excitement as they waited, tractoring in on beams from their eyes. And something else. Relief at his

Matthew Stokoe

presence, the shedding of a burden.

Silence. Silence and open mouths and long lolling tongues thirsty for direction.

What he said could make them his. He looked at what was left of Cripps. Even stripped of flesh the face held the haughty half-smile of vindication it had died with.

Then there were words, hot words, filling him, breathed into his blankness through the cracked bone of a dead man. They came to him free of the worrying filter of the brain, straight up from the gut, instinctive and unexamined, like Cripps' shotgun execution of the fleeing cow.

Beneath him the Guernsey stood proud and still, he could feel its weight pressing against the earth. He scanned the herd and raised his voice.

"What do you want?"

The cows stayed silent.

"What do you want?"

He felt like Mussolini, all jutting chin and cropped head, shouting to black-shirted troops about glory and service.

"You thought you wanted peace. You thought with Cripps dead you could live down here, free at last of man and the horror of your beginnings. But you spent too long waiting for him, you hated him too much. You should have let him live."

The cows rumbled angrily, but the Guernsey trumpeted and they fell grudgingly quiet again.

"You thought he was your problem, but he wasn't. How could his living affect you? You were down here. He didn't hunt you, he only killed what was left behind. You should have watched him more closely, you could have learnt from him.... You thought you wanted your memories – fields of grass, space to chew cud, time for contemplation, all the things every cow before you ever dreamt of. And you could have had it, it's all here – food, safety, silence. But Cripps' death didn't give it to you, did it? It didn't work like you thought because you aren't the same as your memories. You are the first generation of urban cows, different from all others. You were bred for death but you lived, and the old pleasures no longer satisfy. Cripps could have shown you this, he could have helped, he could have made you aware of what you are."

A cow in the crowd shouted:

"Well, the fucker's dead."

Steven climbed to his feet on the back of the Guernsey and spread his arms like Christ.

"But I am here. I have gone beyond him and I will lead you. I will show you where to find the strength to free yourselves of the past. I will show you your nature."

Around him the cows exploded into movement. pushing against each other, butting, colliding, kicking – sides of beef slamming together, spraying wide arcs of foamy sweat. A mad maggot-writhing circle battling against itself.

Steven called down to the Guernsey.

"What's happening?"

"They're trying to understand what you said, see if they're going to accept it. Sit down, man, we might as well split till it's over."

They waited for a break in the churning bodies then raced to dusty calm at the edge of the chamber. Steven dismounted and watched cow madness. He felt strong. His groin burned, the violence his words had sparked excited him – power was so new an experience his brain interpreted the overflow physically.

The Guernsey smirked at him.

"Wild, huh?"

Matthew Stokoe

"Does this always happen?"

"Started after Cripps, like everything else. It's sort of a stampede that goes nowhere. They do it when they think about the future too hard, like it's too much uncertainty for them to handle."

"What happens after?"

"Nothing much."

"Why aren't you with them?"

"I don't need to be."

The Guernsey's eyes were deep and brown and Steven knew that the curl of its lips was a small cow smile.

"Like I said, man, I developed faster. I know some things better than they do. Maybe some things better than you."

"Like?"

"I knew all that killing would change you."

"You said it would fuck me up."

"Well you ain't dead yet, so there's still time. And I knew when you cut Cripps to pieces they wouldn't be able to let you go."

The smile left the animal's face.

"And I know you want something out of it, man. I know you ain't back here 'cause you love us cows so much."

Steven nodded at the herd.

"They're slowing down. How will it go?"

"Don't have much choice, do they? If things carry on like they have been the herd'll burn itself out."

"They'll let me lead them?"

The Guernsey gave a cow shrug.

"Give 'em what they want and they'll follow. We ain't so different from humans."

The cow leant against the wall and small flakes of rotten brick fell within a slower shower of dust to the floor.

"But what I was saying, man. What's in it for you?"

Steven felt his confidence waver just the slightest bit. This animal could make things difficult.

The Guernsey saw his hesitation.

"Don't worry. They ain't dumb, but they ain't as smart as me either. They won't see it so easily. Come on, man, I'll keep your secret."

"Sounds like you want something out of it too."

"Leaders mean hierarchies, and I sure as shit ain't aiming for a place at the bottom. Surprised? Don't tell me it conflicts with this new nature you're about to bestow on us."

The tone was sarcastic and it moved the Guernsey from difficult to dangerous.

"Now, what do you fucking want?"

Steven looked off across the chamber. Most of the cows were still now, slick with their exertions.

"Have you ever seen TV?"

"Jesus, of course."

"Have you ever noticed how perfect life is there? That's what I want."

"All looks the same to me, on TV and off."

"For other people, not for me. But I'm working on it, and wasting time at the plant everyday isn't part of my plan."

"You want money.... It could be worse."

The cows were trooping tiredly towards the edge of the chamber. Steven left the Guernsey smiling to itself and walked out to meet them. While they gathered he searched their faces, and the unease the Guernsey's probing had sparked vanished. These bovine heads wore soft and waiting to be moulded.

A small roan female stepped forward.

"Can you help us?"

Matthew Stokoe

"I can teach you to live with what you have become."

He shouted so they could all hear.

"You have become territorial and aggressive but you refuse to accept it. This is the source of your pain. You need challenge, you need to assert yourselves over others, you need to free what lies within you. I can teach you to do this."

He lifted his arms and the cows knelt before him.

"I can save you."

Later the Guernsey took him to the storm drain.

Chapter Twenty-Six

Lucy was washing clothes in the kitchen sink when he got home. She had become adept at finding and performing chores, endless small charades of domesticity to fill up the hours and rock her brain to sleep. She never left the flat, the horror of other people was too easy to see – the way their faces twisted, the way their backs bent, the uncounted ways of holding themselves and moving and looking at you like they were peeling their heads open to show some pornographic shot of pain. If a woman in a shop ran her hand through her hair in a particular way Lucy would know the agony of her childhood, the terror of her parents, the loneliness and the fear that were now the territories of her existence. So she stayed inside and didn't look and avoided the reminders of what she knew herself still to be.

Sometimes, when Steven was at the plant she went out on the roof and watched the city, but it was meaningless to her. The shapes of the buildings were difficult to focus on, they slipped anonymously away from her gaze into a two dimensional scene that was alien and impossible to interpret, and worse, held no reward for any struggle of understanding. All the buildings were empty.

So she'd go back downstairs and wipe the windows and scrub the concrete in the bathroom, trying to choke back the brainfilth that boiled up against this bland canvas. The time with Steven, with the baby growing slowly in her belly, had been a clamping-down on the constant awareness of the damage she carried within her. The decision to allow the tangling of their lives had provided a veneer of distraction with which she could lightly cover the knowledge that all the systems of her soul and body, progressively corrupted since birth, were still degenerating unstoppably. Before, when she was alone, the dripping accretion of neuroses in the deep pools of her guts was a rain sound across all of life. Steven did not bring the sun, a clearing away of this daily torment – his own goals consumed him too entirely – but he was a separate flow of life, a flow into which she could jump and be carried away from her own, thudding back to shore only when she was too tired to stay away from herself.

In rare, flaring moments of introspection, she toyed with questions of love. But it was a pointless game, made redundant by the need to survive. What did it matter if they loved each other or not as long as each could be used as a screen against the world?

"You're late. I was worried."

"Overtime at the plant."

"Oh."

Lucy served him dinner on the new kitchen table in the freshly painted kitchen. She ate with him. This was all part of it – family, closeness, normality. Eating together, pet names, passing caresses. An illusion of happiness they were both eager to accept and which Steven called real.

"I felt the baby move today."

Steven smiled and got up and put his arms around her from behind, palms flat on her belly, feeling for flutters of the life that would be such an important piece of the future.

"A kid. I can't believe it."

"What's so good about it?"

Steven was a little shocked and went and sat down again.

"What do you mean, what's good about it? If you've got a kid you've got a family like everyone else in the world. You're living like they are."

"It's just another thing to fuck up."

"Don't say that."

"As soon as it's born the poison will start building up in it. Parents destroy their children simply by their presence. And we won't be any different. Shit gets passed on. You can't stop the infection. It seeps through your skin and builds up until it triggers your own shit and pus and then there's no room left inside you for anything else."

Steven reached across the table and held her hand, it felt cold.

"Lucy, you don't have to be like that now. We're in here, we're protected. Poison doesn't grow in this world. I love you...."   .

He had used the words many times in the past weeks to brace her, and he had been sure they were working. But now, for a second, he felt a cold tremor of doubt. Perhaps Lucy was just too far gone.

Then he blinked and breathed and shook himself and the world got itself back in place. He relaxed.

"A child will be good for us, you'll see."

Lucy nodded bravely and tried to smile.

That night he fucked her from behind like she was an

animal. Fucked her and imagined he was riding a cow. After, between the clean sheets of their wide double bed, he held her and they watched neon through a gap in the curtains.

"What happens after?"

"After what?"

"In a year when we have the kid and there isn't any more to do to this place. When we're just living and nothing is new."

"What do you mean? We just keep living. What's wrong with that?"

"It won't be enough. After a while it won't be enough, and we'll have to start moving in the world with everyone else."

"We're happy now and we'll stay happy. The child will make us the same as everyone on TV. It will be enough, believe me."

Lucy pretended to sleep. Steven lay awake, thinking about tunnels under the city, making plans to finance the future.

In the morning, when he rose, Lucy was still asleep. He watched her breathe for a moment, and the doubt from the previous evening returned to hover about him.

He ate in the kitchen then left to meet the Guernsey.

Chapter Twenty-Seven

Underground. Sitting deep in a cool stone passage, alone with the animal lieutenant, listening to ideas filter through beef, catching hints of the desire for power, working on it, weighing payoff and risk, moving further along the road to self-centred cow messiah.

"The place is across town. They're building a skyscraper and its foundations go straight through a tunnel. It's like a huge shell at the moment, just a wide open floor with a little hut stuck way back in a corner. And that hut, man, that's where today each week they get the payroll together. All the work's happening further up right now, so where the hut is, is pretty much deserted. If you want bread it's a place to start."

"Sounds like it."

"Listen, man, this is only temporary, right? We do a few raids until you've got enough money, then you split. OK? You don't belong down here. This is cow country, men can't stay."

"The herd looked happy enough to see me last time."

Matthew Stokoe

"They need to be led by one of their own kind."

"Like you?"

"Yeah."

"There won't be anything to lead if you push me out too soon. They won't take it from you, you're too familiar. They won't believe you can offer them anything they don't already have. Wait. You'll get what you want.... But only after me."

"Don't take too long."

The cow lumbered upright and waited for Steven to climb aboard.

"Let's go see how good this shit you're peddling is."

The herd was gathered and waiting when they entered the chamber. Steven noted with satisfaction the tension that held them silent and still, the hard striations of muscle around shoulders and necks, the dry lips and impatient tongues. He rode tall on the Guernsey, felt himself swelling to fill the chamber, felt sure and strong and knew that each of his actions today would be unquestionably correct.

"Today you begin to learn. Obey me and you will survive. Generations of you will survive to become something more than escaped food. You will cease to be fugitives in the sewers of the city."

Steven breathed the expectant cowstink deep into himself and savoured for a moment its heavy potential.

"Let's RIDE!"

The Guernsey swung into a tunnel and the herd followed. They ran slowly at first – a light canter, legs swinging smoothly forward and back, stretching easily to gather in the ground. Big bodies moving pushed the air and made it whisper. The fall of hooves lay around them in a layer of thick noise, like the sound snow makes when you scrunch it together then run into Mom and your wife all steaming from the cold, into warmth and embraces for ever and ever.

Steven felt his skin glow. He shouted formlessly of triumph and transformation, emptying his lungs. The Guernsey bellowed back and picked up speed and the sound the cows made boiled off ahead of the beef juggernaut and raged against the roots of the city.

At the head of the tunnel that led into the basement vastness of the skyscraper Steven called a halt. The animals panted and strained against the stillness. The speed and the muscle exertion of the race across town had shut down their heads and left them gut-reacting and chafing for action.

He moved them forward slowly. A rough concrete floor stretched off three hundred feet into gloom. Bare, low-watt bulbs hung weakly at long intervals from single strands of flex. At far catty-corners to the tunnel a small Portacabin office had been set up. Yellow light drifted from its windows and collected in a pool around its flimsy prefab walls. The dark shapes of men moved inside.

There were no lectures to give. The cows were on auto and they would do what was necessary to start their healing. And from the rubble Steven would take what he wanted.

"Stay tight and hit hard."

The herd launched itself from the tunnel at high speed. 0-60 in five seconds flat – full muscle contraction. Steven and the Guernsey drifted back from point, allowed the beef mass to swallow them. Wind and cow sweat and body heat. Steven arched backwards and shrieked and lost himself in the madly plunging wave of steak.

Halfway there, the thunder of the stampede brought a man to the open door of the hut. For a second he froze with his mouth open, trying to fit the approaching wall of death into a framework of understanding. Then he shouted and kept on shouting and soon two other men pressed their faces into the space around him. Steven could see their lips moving.

The cows were fifteen feet from impact when the men unlocked enough to think about escape. By then it was too late. The herd hit like a train and the hut exploded in a spray of plastic cladding and moulded fittings. Two of the men exploded along with it. Blood and brain and guts all over the place, staining the chests of the leading animals. The third man was hurled from the wreckage as the cows passed through the ruined structure and ground it to dust. He wasn't seriously hurt and he ran for a flight of iron stairs that led to the floor above. Steven saw it happen and leapt from the Guernsey.

The man moved in slow motion, but Steven was lightning, eating distance so fast he left cartoon streaks in the air. He was a hunter, something loosed to destroy, free of the usual binding brainchatter. No thoughts, only a tremendous sense of himself in the world, and a glorious, instinctive certainty of action.

The man had his foot on the first stair and his hand on the tube steel railing, swinging himself up, when Steven closed with him. He was a big man but there was no possibility of resistance. Steven tore him to pieces with his teeth and fingers.

He bathed himself in blood from the large arteries in the neck while the thrashing body dropped shit down its trouser legs, trying madly to clean itself before God paid attention and took it back.

Steven let him fall face down, put a knee between his shoulders and started to pull back on his head. But before he could get enough pressure on the spine to make it snap, something flickered at the edge of vision and a hoof on the end of a sienna foreleg impacted against the man's face, splintering teeth, caving in the front of the skull.... And stealing the kill.

Steven jerked upright, fists clenched. The Guernsey smirked at him and wiped its hoof absently along the ground to get rid of the blood.

"That was mine."

"Thought you needed a little help, dude."

"Bullshit! You wanted it for yourself."

"Hey, hey." The cow's voice was honey. "I was only trying to help."

"I know what you were doing."

"What's the problem? Isn't this what we're supposed to be learning? Remember, it's part of our nature now."

Steven choked back his anger, this was not the time or the place for a confrontation.

"Let's find what we came for."

The herd was charging wildly around the remains of the Portacabin, playing football with two ragged bags of blood. They were oblivious to anything but the exultation of their new-found power. Steven rode the Guernsey through them and searched the rubble until he found a dented cash box spilling notes into the dust.

"Looks like you scored big."

"It isn't so much."

"You trying to tell me we'll be doing this again?"

"Is that a problem?" Steven nodded at the milling cows. "It doesn't look like it'll be a problem for them."

"Could be, if you push it too far."

Steven caught the edge on the cow's words, but it didn't bother him.

"We'd better go."

The herd was so engrossed in their celebrations that the Guernsey had to ram a couple of them before they took notice and followed. All the way back they trumpeted and bucked and kicked chunks of stone from the tunnel walls. Some of them had hard-ons.

Chapter Twenty-Eight

Back in the chamber, with several thousand in his pockets, Steven watched the cows hurtle the perimeter, their movements stiff with decaying adrenaline. They had tasted a small part of what they might be and the flavour of the drug would not leave them. They moved for no reason but to block out thought because thinking, remembering the liberation of exerting an influence on their world for the first time, was an emotion too potent to bear without some kind of outlet.

Steven felt one more warmly protecting layer wrap itself around his future.

"Bring them over."

The Guernsey moved off without speaking and brought the herd to a halt.

The cows looked at Steven with awe and gratitude, but all he saw were idiot bovine faces slotted into some mechanism, the operation of which he was quickly mastering.

"Have I proved myself?"

The cows yelled like morons at an evangelist

meeting, a chorus of affirmation. Steven waited for silence.

"Do you understand what I have shown you? Do you realise its importance? Your past is dying!"

The cows started bellowing again and Steven had to shout.

"Today was nothing. There is much more.... I will take you beyond yourselves."

Beef surged forward with a thunderclap of joy. Long rough tongues licked his face, his hands, his body, searched under his arms and between his legs. This was cow love, just like the Jesus love that people on TV churned out every Sunday afternoon. Steven let himself float on the warm rasping affection.

Then the tongues stopped and he opened his eyes to the small roan female, arse-on in front of him, offering herself while the rest of the herd watched.

The skin of her vulva was dark brown and leathery, but it was wet too, and Steven knew he had to fuck her. This was a gift that would seal the bargain – cash for him, self-discovery for the cows. He could not refuse. And anyhow, he didn't want to.

He stood on an empty fruit crate that the cows kicked forward, held on to her rump with one hand and rubbed his dick along the oily seam of her cunt with the other. Her glit was thicker than Lucy's and it stuck to him in strands. Dried shit crusted the folds of her arsehole and powdered the insides of her thighs in dusty smears, but the heat that came out of her gash made getting up her the only thing Steven cared about.

When he slid his dick in she shimmied and made small bleating sounds, pushed back against him to get as much as she could. He had to hold her tail out of the way while he pumped. Inside, she felt different to Lucy, the membrane around his dick was tougher and there was a lot more distance. Surprisingly, though, she was quite tight. He fucked her hard, running his hands over the solidness of her flanks, feeling the hair on her hide scrape into the spaces between his fingers. When things got hot and he was really slamming it in, the herd started shouting stuff like *Yeah, do it. Fuck that bitch, man. Do it, do it. Fuck her arse....*

At the end of it, when Steven finished blasting seed, she staggered forward and lay on the ground panting.

The herd applauded.

All except the Guernsey who had remained apart throughout the fucking and who now walked deliberately over and let Steven climb on its back. They left the herd and headed out of the chamber.

"You like cow pussy?"

"Sure."

"Tight enough?"

"Surprisingly."

The Guernsey laughed.

"Must have felt like a cigarette to her."

Steven didn't reply. This cunt cow was becoming a new problem when a lifetime of others had been so recently overcome. He felt suddenly plotted against, as though everything in the world was conspiring to destroy what he had created. He thought of Lucy asleep in bed that morning and how, without the camouflage of domestic activity her madness had seemed so visible.

Were the two of them, Lucy and the Guernsey, circling like sharks, waiting to move in and rip holes in the sides of his life? Either one of them could ruin things for him.

Lucy as wife mother home maker was essential for the continuation of life as he wanted it. A woman was necessary, and all other women were unreachable, out in a world he could never be part of. Only another fuck-up like

Lucy could understand how impossible it was for him to live anywhere but inside one small flat. If she left, or became unacceptably dysfunctional, his home would revert to the shell of its Hagbeast days and the TV would suck back his dream.

And the fucking Guernsey, impatiently counting the seconds until the way was clear for it to unleash its own bovine power hunger. In the flexing of the muscles across its back Steven could feel dissimulation, schemes to take control of the herd boiling like tar under the pale hide. The game would be to acquire sufficient cash and get out before they were put into practice. If the cow moved too soon everything would crumble.

Cunt bastards. In the darkness of the tunnel Steven flipped between anger and fear.

Chapter Twenty-Nine

Lucy was on the bed, legs spread, the insides of her thighs slick with lubricant. She was naked from the waist down and below her swelling belly her hips looked loose, as though she had forgotten that her arse and her legs belonged to her. Steven stood in the doorway of the bedroom, for a moment too frightened to move closer. Beside the bed the endoscope monitor sprayed monochrome static into the room. The probe lay greasy and smeared on the floor.

He sat on the edge of the bed. Lucy shifted slightly, tiredly, watching him with blank eyes.

"What the fuck are you doing?"

He wanted to let his fury overflow, but he kept his voice even.

"It isn't going to work, Steven."

Speaking seemed to drain her.

"What?"

"Why are we together?"

"Because we love each other."

"We're trying to hide inside each other. We called it

love to pretend we were normal but it didn't change anything."

"I do love you."

Steven felt ill. It was hard not to grab her hair and scream into her face I knew you'd do this, you bitch.

But he hadn't known. He'd been frightened of it, but he hadn't known. He had thought her madness equal to his own and that in seeking to flee it she would run the track he laid for her.

· "I thought it would be all right, that if I did what you wanted long enough I might forget about the poison. But it's still there, it's still building up."

She lifted her hands to carve her pain in the air, to make it real enough for Steven to understand, but she saw how useless this was and let them fall limply back across her chest.

"It's still there, Steven."

"I can help you. I want to help you."

"You want to keep this together, that's all."

Her hands fluttered vaguely at the walls.

"Jesus...."

Steven stood up and took a few pointless jerky steps across the floor then stopped and turned around.

"There isn't any poison in you, you're just fucking insane. Look at this place. Don't you like it? Do you want to be back upstairs again, cutting up rats, spending all day thinking how fucked up you are?"

He dragged handfuls of money from his pockets and threw them beside her on the bed.

"Look, and I'll get more and we'll never have to go outside again. The poison will stop, I promise. When the child comes you'll forget. You will, and we'll stay together and be happy."

Steven found himself on his knees by the bed,

fingers curled into a corner of the mattress, tears and spit on his face.

He didn't care about trying to generate love anymore, all he wanted was her, there in the flat, to bear his child, to dress like a wife, to eat with and be warm to touch. As long as she was there and alive he could invest her with whatever qualities he wanted. By sheer effort of will he could make himself see her as he needed her to be.

"You're an idiot to talk about happiness, Steven. We weren't made to be happy. You thought you could be like the people you saw on TV, but you should have looked closer at yourself. You aren't like them, they had lives to build their happiness on, whole backgrounds of normality. You can't do it without that. It's not even worth trying."

Lucy looked grey, exhausted. Her eyes were heavy, her speech sing-song and drifting.

"There can't be any happiness with poison inside, it doesn't matter what you make around you. The only thing you can do is cut it out."

When she started crying Steven lifted her from the bed and held her against his chest. He felt the beginnings of an erection and the return of confidence. Whatever Lucy might babble about things not working, she was obviously incapable of changing them. She was weak in his arms and his snivelling panic of a moment ago vanished as he realised it would be impossible for her to leave him and survive. Like him she needed a construct within which to exist, and she had been so thoroughly sucked into his that to find one of her own again would be beyond her.

He put her back on the bed and while she slept he hung out of the window and stared at the city. It was mauve in the dusk and neon flickered red, blue and green as always, but the spread of the buildings and roads seemed changed, somehow smaller and less significant.

Matthew Stokoe

He imagined the network of tunnels that ran beneath it, imagined cows hurtling through them. His army now. How long would it last? Long enough? He thought so.

When he climbed into bed beside Lucy she felt dead, too heavy and too still. He wanted her to wake and turn towards him muttering soft words of love. But she did not, so he curled the blankets tight about him and closed his eyes. The last thing he heard before he lost consciousness was his own lips whispering *That bastard cow.*

Steven stayed home the next few weeks, watching Lucy and moulding what he saw, in the zero time between eyeball and brain, into something acceptable. She became a non-specific force to which he added the frills of personality, or from which he subtracted those traits that sent shudders through the bedrock of his certainty. He was aware her behaviour was not perfect, but she was alive, she cooked for him, she slept in his bed – he could hold her and warm himself against her. If the nuances of care he had dreamt of were missing, it was something he was prepared to accept.

Lucy operated well enough during this period, but she seldom spoke. She moved heavily through the flat, limbs flaccid until required for some particular task. In moments alone or when Steven sat absorbed before the TV and did not require her interaction, she was motionless and silent, face screwed tight against the knowledge of her deterioration.

One night Steven woke to find her watching tapes of surgery, kneeling, face pressed to the screen as though she could force her head into some other world where the secrets of relief were openly displayed. The doctors were using an instrument that looked like a pair of boltcutters to split the breast bone of a woman with very white skin. There was a lot of blood and one of the nurses had to keep sucking it away with a small hose so the surgeon could see where he was going. When they were satisfied with their hacking they used a clamp to keep the two halves of her ribs spread. The hideous mess visible through the hole reminded Steven of the meat plant and its endless harvest of guts. He watched for a while then went back to sleep. Lucy was still in front of the TV when he woke next morning.

Chapter Thirty

The money wasn't finished, but what he had would not last forever and Steven was anxious to have this last necessity taken care of. He left the flat on a cold clear day that made him think of pine forests.

He felt tired. Transforming Lucy into something bearable was taking more and more energy. And ahead of him, the Guernsey was bound to be a problem. He tried not to think, his will was in danger of becoming diffuse.

There was no cow to guide him from the storm drain, but Steven knew his way.

The air in the tunnels was damp and it made his muscles ache. Sometimes there was half-light, sometimes he had to find his way by touch. And it was in these dark places that his head ran away with itself. The thought of the effort required to again assume leadership of the herd drained him and made his body drag.

At the entrance to the chamber Steven paused and breathed deeply, trying to suck something useful out of the air. All he got was a heavy feculence of dung and animal sweat that further sapped his energy.

He stepped from the tunnel into the copper-glowing vault and found it changed. The hard-packed floor was clean, cow shit piled neatly in far corners. And where last time there had been a chaos of undirected energy, there was now order and peace. A mound of earth had been erected at one end of an avenue of pillars, and ranged out from its base in rough ranks the herd lay at ease – chewing cud, sleeping, nuzzling youngsters, or stood flexing beef muscles, gazing into shadows with far-away eyes.

Above them, on a flat space at the top of the mound, the Guernsey lay with the small roan female. She dozed and oozed recent cow seed from her dark cunt. The Guernsey was awake and alert, watching the herd, and Steven's presence registered immediately. The animal pushed itself upright and tracked his approach, prodding the roan until she woke and cantered obediently down to the herd.

As Steven walked through the rows of cows a murmur spread before him, a shock wave of restrained excitement that rapidly infected the entire herd. Around him cows heaved to their feet, shunting haunches and shoulders out of his way. Tongues flicked out to taste his arms as he passed.

He climbed to the top of the mound. The cows began to bellow and stamp. The Guernsey eyed him closely, gauging strength, weakness, potential threat, then brayed a cow command that silenced the herd.

In the sudden quietness Steven saw that more than the appearance of the chamber had changed. Something in the herd's relationship to the Guernsey had altered also – the way they held their bodies, the angle of their heads, some indefinable rearrangement of muscle and attitude hinted of events during his absence. Steven sensed a realignment of loyalty, or rather a slight splintering of their previous devotion. It was not overtly threatening, but it was there nonetheless and it made him unsure of his position.

"I expected you sooner."

The Guernsey's voice, like his stare, was guarded. Steven scanned the cavern.

"You moved fast."

The Guernsey chuckled softly.

"I told you about hierarchies, man. When there's space at the top it has to get filled"

"I can't see Cripps."

"Yeah, he was getting too much attention, so I moved him. Come on."

The Guernsey led Steven down the back slope of the mound, away from the herd, to a hollow by a wall. Cripps' ragged bones lay in a heap, half covered in cow shit.

"My private dumping ground. Didn't think it was right I should shit with the others."

"What makes you so special?"

"They do, man. After we hit that building site they knew you were right. They had identity again and they believed. But they needed to keep it happening, like they thought it would slip away if they sat around too long. You weren't here, so I stepped in."

Steven felt his skin tighten.

"You took them on another raid?"

"Had to."

The cow looked bland.

"They would have gone crazy. Couldn't let all your good work go to waste, could I?"

The Guernsey was openly mocking him and for a few seconds Steven lost himself in visions of the slaughter room. He came out of it breathing hard and wishing for the balanced weight of a boltgun in his hand. But this was a time to tread carefully, he did not know how far he had been replaced in the eyes of the herd as the bringer of their future.

"How did it go?"

"Fine as wine, dude. Why wouldn't it? How much do you think it takes? They want to be led. They *need* to be led. So we had ourselves a stampede, and I was master of it, man. I mean, I was in control. Found some engineers in one of the sewers. They couldn't believe what they were seeing. They tried to run but the water was too deep. Fuck did they make a mess."

Steven stood silent, imagining the cow in pieces. But some of his uncertainty was leaving him. The Guernsey was too happy as pretender to the throne, too obviously self-seeking. While Steven didn't give a shit about the herd and would gladly use them for his own ends, he understood, where the Guernsey did not, what they needed to evolve as a self-maintaining unit. Like him they had to find or create a new approach to living, and like him they had to unleash something within themselves that would give them the strength to do it. The Guernsey might see that they needed direction, but it could not know as intimately as Steven which way to point them. It saw things only from the outside, it knew the mechanics, but not the reason for their effect.

"You think you've taken my place?"

The Guernsey didn't reply.

On the other side of the mound the herd started to chant, a deep bass rumble that purled up one slope and down the other, like dry-ice fog, lapping across his body in warm waves, calling *Steven, Steven, Steven.*

They wanted him.

He smiled at the Guernsey.

"Doesn't sound like it, does it?"

Steven turned his back on the animal and went up to the top of the mound.

The herd was on its feet, heads raised, throats stretched, throwing his name at the walls of the chamber in hot punches of sound. They were sweating, as though they strained at some invisible barrier, wanting to be near him, to thrust their destiny into his hands.

Looking down on the slick brown backs, Steven felt the return of power. He was equal to the weight of their need. Each successive cry dragged him further from the blurred grey world of weakness that had shrouded him earlier, back into the dazzle of possibility. For seconds on end he saw nothing but the adoration of their eyes. There was a kinship between them. They had shared the release of a kill, and they needed it again as much as he did.

The Guernsey stood close to Steven, its gaze scrabbling insolently over the gathered animals, picking at the shreds of its brief sovereignty. Steven brushed against it and felt hard flesh, this part-time leader was pumping hard to hold on to the advances it had made during his absence. Conflict was inevitable, but it would come later. The Guernsey was not stupid and a wrong move now, in front of the herd, would destroy any future chance it might have at leadership.

The cows were getting jittery, they wanted action. Steven felt their tension in himself. He flexed the long muscles of his thighs and arms, and every cell in his body opened up and started screaming for the heroin of concentration – that state where the only thing that exists in the whole world is the thing before you, in your hands, bleeding and dying.

He whispered close to the Guernsey's ear.

"What was next?"

"Huh?"

"Your next raid?"

"Another raid? Sure, man. Have to keep things rolling

along."

"Well, what was it going to be?"

The Guernsey sighed.

"An underground station. End of the line. There's a ramp from the tunnel to the platform and late at night not too many people... but enough. Thought the great leader would have a plan of his own."

"What about cash? There won't be much in a station."

The cow laughed at him.

"We don't need money, man. Are you blind? We're fucking cows."

It nodded at the herd.

"They want what you promised, not some mercenary substitute. If you can't handle that you're welcome to fuck off."

Steven didn't have to think about it. He needed money, but he needed other things as well. He needed to feed his desire for power and the love of the herd.

"It'll do."

He stepped away from the Guernsey and spread his arms to quiet the chanting that had become almost deafening in the stone chamber. There was silence at the instant of his gesture.

"I am back to lead you."

The cows shrieked.

"There are things still to learn," he looked sideways at the Guernsey, "that I alone can teach."

The cows shivered and rocked and screamed:

"Show us."

"I will show you."

Steven's words came back at him fast. They snagged on his hot skin and burrowed in, infecting him. He believed what he was saying, he really did want to show them. He wanted to be flying through tunnels with them and fuck everything else, heading for the bliss of a focusing sensory overload – Cripps communion.

He swung up on to the Guernsey, oblivious to the animal's flinching, and charged down the mound, straight at the herd, then through them, howling for them to follow. Spit streaked back from the corners of his mouth.

To Steven, with the thunder of the herd behind and the muscle play of the beast beneath him, the journey to the underground station was an ecstasy. He kept no check on the Guernsey, he cared nothing to follow progress or determine location. All he wanted was movement towards some sense-sucking activity – towards strength and away from weakness. The Guernsey would run true, it had no choice, the herd would allow nothing else.

Time passed as fast as the cows could force it to. Then they were stopped in a tunnel, in the shadows twenty yards back from where it widened into a station. Double tracks of bright steel fired straight from their feet, out past the platform and back into another tunnel on the other side. For a second their shine held Steven's gaze, drawing him out through his eyes until he felt he could stand, one foot on each rail, and slam out into the almost happening bloodbath like some angel on a rocket sled, sure and unstoppable. And on and on, beyond now, through every future experience he might have, on an exact course where no mistakes were possible....

Then the Guernsey started to whisper a plan of attack.

"OK, first section move along the tracks, right by the platform. Go slow, let 'em see you. Then section two, while everyone's looking at them, take that ramp there, stay quiet until you get level, then steam in and waste the fuckers. Some of them are going to fall off the platform on to the rails, behind section one. So section three, that's your job – sweep up behind and stomp anyone there...."

Steven only half listened, the Guernsey's voice fuzzed. What it said was so irrelevant he could not spare the energy to follow it. Instead, he listened to the crunching of the rail bed as the cows shifted weight, the soft snorting and farting of odd animals, the scratching of a rat back down the tunnel.

He tore his eyes from the rails and looked out into the sizzling brightness of the station – no trains, a ramp, about twelve people waiting to get back to homes and wives and children.

The Guernsey was fucking things up. It was planning, rationalising, making clinical what should have been wanton and intuitive. It had missed the point – they needed recklessness, not precision. Yak, yak, yak. All the words were wrong.

So.... Steven sat up high, sucked in his breath, twisted both the Guernsey's ears and shrieked over its interminable babble:

"Fuck it all. CHARGE!"

And yanking on the ears, forced the Guernsey out into the white light of the station. The herd, in an anxiety-ridden limbo between their past and their future, plunged after him.

From shadow to visibility.

Steven released the Guernsey's ears and the animal, knowing it was too late to rebel, shot up the concrete ramp and on to the platform like an insane three-ton truck. Close behind it the herd was a boiling collage of drawn-back lips, red eyes and pistoning limbs.

The people on the platform looked up from their feet and their newspapers and pissed down their legs.

None of them fell on the rails. Some started to move

towards the exits, but there just wasn't time. The herd hit them in a broad wedge, impacting against one then another and another, not stopping but carrying the bodies forward in a bleeding collected mass.

Steven watched the white-tiled wall at the end of the platform come towards him as though he ordered it to move. He shouted, and the sound he made was the sound of the herd – hoof clatter, lung bellow, muscle noise. Blood and shit blew back across his face and there was nothing in the world but its wet salt taste.

The herd was around him, pressing close, jostling to be in front line contact with the moaning tangle of bodies. The Guernsey's head was soaked with blood, part of its snout buried in the burst abdomen of a young woman held sideways and aloft by the snouts of other cows. Steven looked closely at her and saw her eyelids flutter. Wall impact was seconds away and there was nothing, absolutely nothing on earth she could do to escape dying under an avalanche of beef.

Involvement on a more personal level was imperative. He reached forward, gripping the Guernsey with his knees, and took hold of her head, thumbs gently closing her eyes and resting there. He locked his arms straight and drew his fingers back until they were curled behind the woman's ears.

Her head was the first thing to hit and the impact drove Steven's thumbs through her eyes and into her skull. Viscous milky sludge squirted from her sockets and wet his forearms, then her head burst in his grasp and fountained blood and brain against the tiles in an enormous ink-blot pattern.

The herd broke against the wall, driving the bodies into it, trying to turn to avoid it themselves but failing and staggering and slamming into each other in a grunting cow pile-up. Steven was hurled from the Guernsey but the ruptured carcasses of the dead train travellers cushioned his fall. Cows threw themselves down around him and made a cordon against the shunting rear ranks.

For long floating seconds Steven lay where he was and watched the colliding animals, feeling other people's blood soak the back of his clothes. He was safe and warm and free of doubt, recharged by murder.

When the inertia of the stampede was spent, the cows picked themselves up and stood in a dripping curve before him, waiting for some indication of what to do next. Off to one side the Guernsey was alone, treading on a body and turning it to paste.

Steven rose and touched the red-stained heads of those closest. They nuzzled his palm gratefully, but their eyes strayed to the dead people behind him. He lingered on the small roan female, stroking the underside of her chin.

The herd quivered impatiently.

"What do you want? Right now, without thinking, what do you want?"

When she answered her voice was loud and Steven knew she spoke for them all.

"I want to know always that I was able to do this. I want to carry the smell of their blood deep in my hide."

The cows shouted agreement and Steven shouted with them, goading, exhorting, whipping them into a frenzy. When a gap came in the noise he told them to do what they wanted to do and jumped safe to an angle of walls away from the pile of bodies.

The herd fell on the heap of flesh, smearing blood and piss and excrement over as much of themselves as they could. They rolled in the mess and drove it through hair to deep layers of skin.

Matthew Stokoe

Steven watched as bodies were rent. This wild fulfilment of cow desire was a reflection of the power he felt raging within himself.

A presence at his side – the Guernsey, face, chest and forelegs bloody, but uninvolved in the platform brawling. This animal's own bloodletting had occurred before the sanction that had freed the rest of the herd to action. The implication was not lost on Steven – for this one his permission was of little consequence.

"You think they needed you for this shit? Jesus, the men I led them to in the sewer weren't any different. They like you, but they'll find out you're the same as Cripps and one day soon you'll be history. Hear what I'm saying? I can lead them just as well as you."

On the platform the herd, exhausted now, and sticky with blood and pieces of flesh, were lying down.

"A few raids to get me some cash and they're yours."

"Don't pull my dick. You're getting into this killing thing. You got Cripps inside you and he's waking up. You're a bad man."

"How can you know anything about me? I'm a human, you fucking animal."

The Guernsey's eyes narrowed. Things were out in the open now. Each was a threat to the other, and each knew it. Steven felt no fear, though, the thrill of the woman's head exploding in his hands was too fresh.

"There's a train around this time, better find someone else to ride, motherfucker."

The animal moved off to rouse the cows from their dozing. They obeyed its commands but their eyes flicked to Steven. When they were ranked and ready to move and he still had not joined them, the roan female broke from the group and approached him.

"It's time to go, there's a train coming."

"I know."

"I will carry you."

Steven stroked her head, scratching gently in the thick curling hair that grew like Persian lamb between her ears. Then he called to the herd.

"I will follow. Go."

The Guernsey bellowed and for an instant the herd froze to absolute silence. Striations of locked muscle cast light webs of shadow across shoulders and legs, full capacity lungs bled oxygen into frothing blood and in every animal brain switches clicked to action mode. Time started again with a snap and they rocketed along the platform, swirling litter into mini tornadoes with the wind of their passing.

Steven watched them disappear into the dark oval of the tunnel mouth then moved to the carpet of plastered flesh and began peeling pieces from it. He found a T-shirt still clinging to something that looked like a lung and used it as a sack. He chose meat that looked soft.

The roan was getting nervous and somewhere, far off, the whistling roar of a train began to grow. She whinnied, but Steven pushed it to the last few seconds, stuffing his bag full, collecting meat until the dim glow of the coming train's lights fanned against the black curve of the far tunnel.

Then they ran, Steven at her side for a few seconds then swinging up, pony-express style, on to her back. His makeshift sack was heavy and it thumped against his thigh. Behind them the train blundered into the station and sent hissing clicks along the rails under their feet. The roan flattened her ears and stretched into a gallop, surefooted on the sleepers and gravel, straining to do service to the man she carried. Time flowed perfectly, it opened up and made space between them and the danger behind.

Chapter Thirty-One

The roan skidded sideways in a shower of small stones and they slid into the damp quiet of an intersecting passage that had led them to the underground line earlier that night. She paced on a few yards, winding down, catching breath, then stopped, sweating, dull-gleaming in the dim light. Steven dismounted but stayed close, running his hands along her sides, forcing out small rills of perspiration with the edge of his palm. Her damp heaviness excited him and his cock went hard.

"We're safe now."

Behind her. She shook her arse. He used both hands to stroke the insides of her thighs, exploring the curves of muscle, bouncing his fingers over veins brought to the surface by her running. Up to the dark folds between her hips. Then into her vulva, parting it with his thumbs, bending close to catch its woody smell, pressing his mouth against it and swallowing what he found there. He ran short dreams of pushing his head all the way inside.

But he didn't want to hide himself in her. He wanted to command and control her, to slam himself in like his cock was a weapon, again and again until something ruptured or they both blacked out. He wanted to pump her so full of come that she burst.

So he made a pile of stones and climbed up on it and fucked the living shit out of her. She didn't resist because she wanted it as much as he did and their moans tumbled along the walls and spread out under the city in a vicious cadence of dominance and submission.

Chapter Thirty-Two

In the chamber the Guernsey was back on its hill. The herd was agitated, moving restlessly in knots of five or six that wove in on themselves, tightening until they fragmented and split to join other knots. The Cows had emerged from their killing session unhappy. The satisfaction of the station had dimmed during the rush back to the central chamber and they were left wanting some undefined further experience to complete their journey towards self-realisation.

Steven slid from the roan like an indian, flipping one leg over her neck and landing with knees flexed. He shouldered his bag of human scraps and strode through the milling cows to the base of the mound. The Guernsey watched him hatefully.

He waited on flat ground, letting seconds pass. Above him the Guernsey stood set and heavy, gathering menace, as though with enough of it he could build a wall around Steven and blot him out. Finally the animal spoke.

"You failed us, man. You didn't bring us more than we already had. You're a nigger down here, you don't belong. The herd should be led by one of its own."

Steven rested the bag of meat against his foot.

"You can't do it without me. You haven't learnt to carry your killing inside yourselves."

"Early times yet, man."

"Time won't do it. What is it that you think I have shown you?"

"That we can kill men same as they kill us."

Steven laughed across the words.

"I am not teaching you to destroy men, but to become like them."

He lifted the sack of flesh above his head and addressed the herd.

"Here is my final gift. Here is the last thing you need to escape the weakness that binds you to your past. It will leave you no choice but to seek experiences that free you from yourselves."

He raised his voice.

"When you stampede are you not free?"

They bellowed.

"When you feel bones break and flesh tear beneath your hooves are you not living as you wish to live? Strong and free of the uncertainty that has dogged you since the death of Cripps?"

The cows bucked and yelled and tossed their heads. Steven had them. He saw TV footage of Nuremberg and thousands holding flaming torches, heard chants that thundered in a cold Aryan sky. These animals were food on four legs, not blond storm troopers, but it was the same. And he could deliver.

"Come forward and be blessed."

Before the first cow could move the Guernsey pounded down from the top of the mound and planted itself in front of Steven. Its eyes were bloodshot and there were deposits of dry spit in the corners of its mouth. It stood there, grinding its teeth, twitching, pushing against the line that marked the end of antagonism and the beginning of violence, wanting to cross over and crush Steven flat. But it did nothing, just breathed until its leg muscles relaxed enough for it to strut to the side of the mound and look on malevolently.

Steven reached into his bloody sack and took out a small piece of flesh.

"Who will be the first?"

The roan moved close and ate from his hand. She kissed him with lips stained red from the meat and moved away. Then the others came.

So it went. Cows filing by in the gloom of the cavern, taking their taste of a food that would drag them into their future.

Back in the shadows the stream made wet noises and the rock of the walls ticked with the strain of hiding these fast becoming dangerous animals from the city.

The Guernsey was the only one who would not eat.

Chapter Thirty-Three

Lucy lay on the floor. The flat was empty, the windows were open and it was cold. She looked down at herself, past her breasts to the dome of her belly. She couldn't see the hair on her cunt because it got in the way. Before he left that morning, Steven said he could feel the child kicking. But it wasn't a child, she knew that.

In the months since her periods stopped she had pretended she believed what Steven told her, had let him believe the thing she was carrying was his. But today she stopped pretending. Today, flat on her back with her skin goose bumping and tightening around the thing in her belly, she let herself see it for what it was – a hard black stone of poison that had grown and grown until her body had to stretch to hold it. And the fucking thing was going to keep growing until it killed her.

Earlier that morning, squatting naked, she had tried to force her hand up through her cunt and into her belly. She used lubricant and corkscrewed her fingers until spots of blood fell on the new lino between her feet, but she couldn't get further than her knuckles. It was after that that she had lain down on the floor.

She opened her wallet of scalpels, buried at the back of a drawer since the move to Steven's, and cut slits in her cunt, up through the clitoris and down almost to her anus. When she woke from a few moments of unconsciousness her neck and breasts were spattered with vomit, but pain was a small price to pay for the removal of poison.

Her hand went in much easier this time, but even with the new looseness and the slipperiness of blood she couldn't force it in much past her wrist. The angle was awkward and the inside of her forearm jammed against her slashed clit. She groped with her fingers but she couldn't reach anything.

The pain from her cunt spread out like acid over her thighs and pelvis, but the stone inside felt worse. She reached for the scalpel again.

When she made the long grinning incision around the base of her stomach things started to get cloudy. Her arse felt like it was floating half an inch above the floor, bumping softly up and down. There was an awful lot of blood and somehow it must have got in her eyes because everything was hazed red. It was a kind colour and it wanted her to sleep, it lay on her arms and made them heavy to move. But there was something she had to do, something under the blanketing pain that couldn't be ignored.

She dropped the scalpel and shoved herself up on to her elbows until she could see the gouting slice across her guts. The red lips parted as she moved and she was happy that her body was open to her at last.

Her hand moved surely through the cut skin into the wet heat of her womb. She felt it immediately. A hard thing, a thing of solid form that no-one could call an imagining. She smiled to herself and closed her fingers around it. It was oddly shaped – she had expected something smooth and oval – and it was slightly rubbery. But it was there and that was enough.

The red was thicker in her eyes now and she was weak, so weak she had to lie back on the floor. She took deep breaths and braced herself, gathering strength. She couldn't tell if her eyes were open or closed, but it didn't matter because she was going to do it. She was going to empty herself of a lifetime of pus.

She made sure her grip was tight, sucked in one last lungful of air and dragged the thing out through her wound. It was too heavy for her to hold so she dropped it on the floor. Her eyes had ceased to function but she pictured it there, lying black and stinking, and felt waves of relief wash away her pain, taking her away from herself into the soft darkness of freedom.

She died feeling clean, the happy weight of the poison resting outside her, pressing against her hip.

Chapter Thirty-Four

In the streets, high up on the surface of the city, Steven moved through a twilight of scattered stars – cars, lights, people, all of them carried a warm nimbus of intensity that illuminated but did not threaten. Everything was safe now, the herd and its potential as a mechanism to fund life was his. The sacrament of human meat had zapped cow cells and completed a mutation that had started with their escape underground. They were changed and they were happy with the change. The tearing anxiety of the intermediate period was over and they knew themselves for what they were – beasts with drives and the ability to satisfy them. They were no longer the frightened, hiding property of men. They had become hunters, able to abandon themselves to any action that advanced their well-being. Their future was cast in iron and their gratitude would be as enduring. He was God to them now, the giver of life, for they would have perished without him. The price they paid for survival, if it was a price, was an ongoing hunger for human flesh.

In his street the sky was clear and the sodium

vapour lamps burnt like suns. The shadows they cast were pure and sharp.

He climbed the stairs to his flat, picturing Lucy's cunt spread warm and waiting just for him, and afterwards the drifting hours in bed next to her before she rose to feed him. TV side by side, time passing without danger because of her presence, arms around him, fingers on his skin... warmth, comfort, safety.

But there was no warmth when he entered the flat.

He found Lucy in a lake of blood on the kitchen floor, a hole two inches above the tuft of her cunt and a yellow foetus corpse snuggled against the outside of her thigh.

He started to puke but it died in his throat and his stomach went still and cold. Lucy was gone and she had taken everything that only a moment before had seemed so unassailable. Without her he would be alone – no soft body to lose himself in, no breasts to mother him, no movement in other rooms as he slept or watched TV. Only emptiness.

The cold spread out from his guts to the walls of the flat, icing them, freezing the recently applied paint, dulling colour and texture. It devoured even the light. Peripherally he saw the skeleton of the flat show through, a superimposition of the place as it had been before the Hagbeast left.

Every night spent alone and frightened in the creeping dampness of his room hit him in an overwhelming pulse of memory. He fell to his knees under its weight, crying out against the fear of having to return to such desolation. No wife, no child, no perfect family. Not even a dog this time round.

He stirred the blood on the floor, it was thick and clots of it stuck to his fingers. His tears made milky-pink splashes on its congealing surface.

Matthew Stokoe

Thieving brainless bitch. Stealing his new life and taking it with her, over into the dead space where the Hagbeast waited cackling for him. It wasn't fair that after a lifetime of pain such a small thing as happiness could be taken away from him by one mad cunt.

He kicked at her head for a while, but the impact of his feet didn't change anything.

He lifted the rubbery foetus corpse from its sticky bed beside Lucy's leg and nailed it to the wall at head height, this thing that was almost all head, with a couple of pointy knives from the kitchen drawer. It didn't look like Jesus. It looked like what it was – dead, deformed hope. It was never going to grow sandy blond hair or wear faded dungarees, it was never going to play in corn fields, and it was never, ever, going to be something to love. On the wall it was a badge that screamed *Idiot, sucker, deaf dumb and blind motherfucker. Did you really think you'd get more than this?*

Steven ran away from it, out into the hall and down to his room... into the room that had held him in its grim arms through all those long nights when the Beast had raged in the corridor outside. The new paint and the fresh fabric meant nothing, the walls that had watched over his horror were there still, under all of it, and they closed their old darkness about him like a revolting but familiar narcotic.

He turned the TV on, but he lay on the bed with his back to it because it was a liar. It held up pictures and said you could be like them, but it didn't tell you how easily everything fell to pieces.

Night wore on and the TV smeared the walls with its deceitful colours. Steven hunched into himself on top of the blankets and thought of nothing.

Chapter Thirty-Five

Morning broke mercilessly and everything was the same. Steven woke but didn't move, just stared at the wall, blinking when he had to, unaware of his body.

Programs rolled through their schedule, hour and half hour changes. Morning TV, talk shows, quiz shows... on into lunchtime soaps and movie matinees. The sun strengthened but the temperature in the room did not rise. Steven watched the wall brighten but it meant nothing. The passing of time meant nothing. There was nothing ahead or behind, to look forward to or back on. No reason to move or feel. So he lay and stared at the wall and made no effort to interpret the scrabbling TV noise.

He stayed this way for three days, blank and adrift. His body pissed and shitted for him but he did not feel the wetness or the hard pellets that flattened between his arse and the inside of his pants.

And all the time the TV babble ate closer to a part of his brain that would listen, ate across the wasteland of his shock to the last soft collection of cells that might react.

On the fourth day Steven heard what it was saying,

heard words and decoded them into meaning, began idly to listen to the perfect short sentences of commercials. Then smelt for the first time the stink of his own filth, felt it itching in the grooves of his body. And another smell that was worse and came from somewhere outside the bedroom.

Slowly, achingly, he rolled off the bed and stood. His back felt broken, he had neither the strength nor the will to straighten fully. He started for the bathroom with the dim notion of cleaning himself and had to force every step. Motion was a battle against a lethargy that hung from his shoulders like a cloak of chains.

Naked body – shrunken, water-logged penis, dark shit smears. He stepped into the shower and leant in a corner so he wouldn't fall.

Into the hall without bothering to dry himself, dripping, trundling along like some sleepwalking Frankenstein.

Lucy was ripe in the kitchen. Dark skin, bloated, heavy, like she had never lived. Lying there in a cracking ice-rink of dried blood. The smell was appalling, Steven sucked it in to see how much he could bear. With his eyes closed it felt like he was standing on the edge of a rotting, canyon-wide cunt, about to fall in and be consumed by its geysering glit.

When he tore her free of the blood she bumped across the floor like a piece of furniture. He dragged her by her ankles and her arms caught on the legs of chairs.

Getting her to the roof was hard but Steven suffered it like a mule. The physical pain of lugging her up the ladder, the frustration of trying to fit her around corners, was just another part of unbearable existence

There was still something of Dog wedged between the chimneys. The upper part of its body was mostly stripped, but the lower half, beyond the reach of birds, retained a covering of dried meat. Steven peeled the carcass from the brickwork and carried it carefully to where he had placed Lucy.

They had never known each other, these two dead things, but he had used them both for love and it seemed right that they should be together. He picked away the scabbed blood that sealed Lucy's wound and pushed Dog's body into the hole that had been home for his dead child. He couldn't fit it all in and a small bundle of bone ends stuck out between the bruised flaps of skin, but it was the best he could do. He left them both for the crows and went back downstairs without bothering to look at the city.

It took him several hours to clean the kitchen. The blood on the floor and the foetus hanging from its rack of knives did not bother him. Nothing in the flat was worse than anything else. Blood or lino or wallpaper or paint, all formed equally a cage for his pain. But getting rid of the visceral litter filled time, burned off some of the minutes between now and death in a blank monotony of physical action.

Later it started to rain. Steven sat on his bed and stared blindly at the TV. The sound of falling water closed off thought, leaving only a dull resignation to loss.

He sat there for days. And outside the rain never stopped. It came down in thick rippled sheets, soaking the old red bricks of the building until they became soft and started to flake, and the cement, already weak from resisting the black cancer of sadness within, started to dissolve.

The structure got heavier, sick with its own weight. At night it groaned and its foundations shifted in the mud. Until one dawn, in the chain of pouring dawns that had tied Steven to his bed, the entire back wall of the apartment house pulled away and slid, slow motion heavy, into a pile of smashed brick that bled red pigment into the flooded trash of the yard.

The roar of falling masonry tore through the mist of Steven's misery and announced itself as something urgent that required attention. So much noise meant the world was breaking in, and he could not tolerate that.

He went to the back of the flat, down the corridor and into the Hagbeast's old room. And stopped dead, aghast at the completeness of his nakedness.

Where there had been a wall and a window the rain had left an open square, empty of everything but a dull, drizzled view of warehouses and flooding streets. As though God had reached down with a knife and neatly cross-sectioned the house, looking for sinners.

Steven edged to the hole and leaned out to look at the floors above and below – sad dead rooms full of rotted carpet and the refuse of abandonment.

He hung there in the gritty rain, tempted to laugh at this final ravaging of hope. How absurd it seemed now to have dreamt, to have dared to want a mapped and selfish piece of happiness. All the wishing, all the insanely unexpected aligning of factors – Lucy, Cripps, the Hagbeast's death – all as nothing now, exploding out over an alien cityscape through a hole in his flat, like air into a void.

It almost turned him to dust.

He went back to his room and shivered. The city would not let him remain in so dangerous a place for long. There would be men and trucks, officials assessing damage and cost, questions, forms. They would reach in for him like a giant hand, the sharp end of society wanting to know how it happened, squeezing him for information until he was destroyed by an unbearable level of contact with a world he longed for but was unable to enter.

They came early. The rain had lightened to a fine spray but

the sky behind the tower blocks in the east still looked heavy. Steven watched from the kitchen window, saw the shiny fire trucks pull up, looking foreign and too clean in the filth of the street. Counted the police and the men who stepped from bland social service cars, imagined their lives and envied them.

When they reached the street door and started pounding he crept to the hole at the back of the flat and climbed down a pipe to the yard. There was no reason to look back, but he did. His one-time refuge, site of extended horror and brief happiness, was a stack of gaping boxes.

The terror of no place to crawl back to swooped down on him like a hawk and he puked into a puddle. He had to be moving, to get away from this magnet for people. The sodden wooden fence at the end of the yard was too weak to stop him and he moved out into the city.

Matthew Stokoe

Chapter Thirty-Six

He walked fast in the sick light of early morning, through a maze of drab buildings that covered the area like some anonymous infection. The streets were empty, but they wouldn't stay that way. In apartments and houses people were stirring – rolling out of bed, dressing, eating breakfast, getting ready to race out and crush him with their zest and their interest in life.

The rain stopped and he ran. His thoughts splintered against the problem of how to protect himself from this coming wave of contact. He knew he was weak again, that he had lost the ability to direct himself somewhere back in the days after finding Lucy dead. But this thought and all his others would not hold still to be dealt with, they rattled through his head in bright streaks of fire, building to a hot ball of anxiety that prevented the making of any decision. He ran faster, pumping his arms, throwing his head back and slitting his eyes.

There were people on the streets now – a lone factory hand with a bag of lunch, then two more, then further along a cluster of process workers, a truck driver stepping from an early morning café, still chewing. More and more of them, clogging the sidewalks. The faster Steven ran, the denser they became. His head snapped left and right, desperate eyes searching for a hole, an alley, a crack in a wall, anywhere he could be alone and safe. But the street stretched ahead unbroken, built solid on each side with factories and small grubby shops that served the workers.

Each person he skirted, every group he dodged, was surrounded by some vicious field that corroded in successive layers whatever force held him together as a person. He could feel himself disintegrating like meat in a bath of acid. Soon nothing would be left to bind the separate parts of himself and he would splurge along the sidewalk in a bleeding pulp of guts and degenerating tissue.

He was lost. The street seemed endless. He ran until the soft insides of his lungs felt scorched and the muscles of his legs began to cramp. He ran without hope of escape, filling the world with the noise of his pleading, until, between the loading yard of a chemicals storehouse and a diesel garage, he found a blessedly deserted alley.

Into it – sobbing, sprinting far away from the growing crowds of morning workers, far away into an ecstasy of solitude, past trash cans and rear entrances, over straggling weeds that poked through cracks in the tarmac up close to the walls. The alley turned sharply to the right, so sharply that from a distance it looked like a dead end. Steven let his tearing eyes close and coasted round the corner, losing speed, gulping air, slowing, slowing, relaxing, stopping, bent forward hands on knees, breathing, trying to think.

For a moment there was no sound but the rasp of his throat and the thud of blood in his head. He let the red darkness of his closed lids calm him.... Then he heard it, all

　　　　　　　　　　　　　　　Matthew Stokoe

about him, in front, behind, on either side – the sound of bodies moving, people talking, feet on ground. He straightened slowly and opened his eyes.... And froze.

The alley had betrayed him. It had emptied him on to one of the busiest thoroughfares in the city. Ranks of people broke against him, split and reformed around him, cursed and moved on.

Steven pissed himself, held his head in his hands and screamed.

Chapter Thirty-Seven

He woke in a narrow concrete tunnel, hunched into the small space between its smooth walls, knees against chin, arse in a six inch stream of dirty water. Tired light worked its way through a grate immediately above his head and thin images of memory played against the bars – the fall from the sidewalk into the gutter, the mad clawing at the cast-iron of the drain, people stopping to laugh as he finally pulled it aside and fell head first into its dim protection, scurrying like a rat further into safety. Then a dreadful weakness, and darkness.

It was early evening, Steven was cold and wet and hungry. He had nothing to go back to – no wife, no flat, no comforting lair, and in the gathering dusk his snivelling echoed through distances of tunnel.

He slept briefly, motionless as a dead thing in the city's current of piss. And when he woke, his mind, frantic for succour, sought the only source of affection left to him. It streaked deep under the city and brought him back the image of a cow.

Matthew Stokoe

For hours the tunnels were unfamiliar to him. He crawled until they became large enough to stand, and then he walked. He waded randomly through sewers, swam artesian streams, dragged himself through cracks in the earth of the city. Until, eventually, he began to move more surely, making turns and changes of direction without thinking, but knowing they were correct, drawn by some lodestone of other living bodies, bodies that would accept him. Surrogate mother love, heifer sex, bovine companionship.

And then it came, as he knew it would – the first brick, the first yard, the first curve of stone that he recognised.

Above, the city churned with its small-hour pleasures, but for Steven it had ceased to exist. There had never been a home or a woman or a dog. There had been no TV, no nights, no days, no growing, no wishing for another life. All there had ever been were these tunnels and the herd at the end of them.

He moved faster, needing to be there, to be back at the heart.

Chapter Thirty-Eight

Close now. Soon he would be among them, drawing love from them like a black hole, drawing their companionship into coils of protection around him.

He smelt them first, the musk of their dung and sweat, and then he heard them. Not at rest, but stampeding, racing towards him in an explosion of sound. Closing, too fast to stop. Not intending to stop, thinking only of speed and momentum.

Steven jumped for an alcove as they slammed into his stretch of tunnel. And watched them pass, a mad conglomeration of horns and eyeballs and locomoting bodies, too caught in the excitement of stampede to notice him. He could taste their immersion and he wanted to be with them, lost in the pureness of their motion, safe from the bladder-weakening anxieties that had chased him down here.

The Guernsey was leading. Cunt.

Then a swallowing up of sound, receding hoof fall... and silence.

Matthew Stokoe

The chamber was empty, no-one missed a stampede. But they would be back, smeared red and sated, the addiction to human flesh that Steven had begun reinforced by their latest dose.

Meanwhile, the stream still flowed and the Guernsey's mound still stood. At the centre of cow sanctuary Steven breathed deep.... Yes, this place could be home.

Behind the mound, under a crusting of dry cow shit, Cripps' bones were hard and smooth. Steven squatted beside them, pushed his hand through the ivory tangle to what desiccated organ tissue remained and closed his fingers about it as though it were a soul. He stayed there a long time with his eyes closed, not sure what he was doing, but remembering the horrors that had become glories through the conduit of Cripps' madness.

When he opened his eyes again he wanted something heavy and dangerous to hold, so he pulled Cripps' thigh bone from its socket. It had been broken just above the knee and one end was sharp as a spear. He weighed it in his hands and it felt good.

On top of the mound he played back scenes of his last time there – the upturned cow faces, the adoration, their surrender to his guidance. The chamber was open before him, waiting for him to fill it, daring him to take its emptiness and fashion himself a womb. He had done it before, the cows had arranged themselves readily enough to his words, surely he could do it again.

But so much rested on it now. He no longer had Lucy and the flat to fall back on if he failed.

He buried himself at the back of the mound, beneath Cripps' bones and the Guernsey's shit, and waited.

They were a long time coming and he slept.

The ground was shaking when he woke. Herd returning, circling the chamber, decelerating, dropping human bodies in front of the mound and gathering close. Steven listened to their panting and their short impatient movements from beneath his crunching blanket of excrement. He could see only the top of the mound where the Guernsey stood, planted like a cannon, surveying the herd.

"I have led you to bounty again. I have given you the meat of survival. EAT!"

Steven twisted Cripps' thigh bone in his hands. From the other side of the mound came the sound of flesh and cartilage being torn – bovine mouths chomping and slobbering – but it could not drown the roar of anticipation that filled his head. How good it would feel to waste that pompous thieving fuck.

After a while the herd quietened and dozed. The small roan joined the Guernsey on top of the mound and took two feet of its ribbed cock from behind. Steven watched the black length slide in and felt more keenly than ever the need to be close to some other living thing, to have someone accept him and provide comfort.

He closed his eyes and watched his mother choking on shit, saw Cripps flayed and bleeding, a woman's head exploding against an underground station wall – pumping himself up as the herd slept.

When there had been no sound from the herd for an hour, Steven rose carefully from his bed of shit, flexed his arms and crept with the thigh bone to the top of the mound. The roan was asleep. The Guernsey, facing out over the sleeping ranks of the herd, was not. Steven stepped quickly forward and squatted, the animal jerked in surprise.

"Fuck!"

"How's it hanging, Mein Fuhrer?"

"You made a bad mistake coming back, dude."

"Didn't have a choice."

"Too bad, there ain't room for you anymore. I've consolidated."

Steven felt the familiar burn of an approaching kill, the rise of some chemical from the guts that doubled strength and crystallised thought. And when the cow started to push itself to its feet he knew it was time to act.

At the edge of his field of vision he saw the roan lift her head and open her eyes. She looked at him with love and he knew she wanted him to be the winner.

The Guernsey had its hind legs straightened and was about to shunt the front of its body off its knees. Steven tensed in his crouch for a second, then leapt upright, holding the bone with both hands, angled and firm. The splintered end tore easily through soft beige neck skin, ramming in and cutting a vein. The Guernsey made a choking noise and lurched forward. Its chin hit the ground. Steven tried to tug the bone free, but it was stuck too deep in flesh and he had to brace his foot against the cow's cheek and lever it out. A single thick stream of blood slopped from the hole it left.

The Guernsey jerked around trying to lift its head and pull breath into its body. Blood mixed with snot pooled around its muzzle. Steven felt like a god. He struck again, down through the top of the neck, just behind the skull. It was easier this way and the point of Cripps' thigh bone made it out into air on the other side.

He missed the spine but the Guernsey was so badly damaged, was struggling against so much intrusion, that its hind legs buckled. Steven laughed and straddled its back and drove the bone in again and again until he was wet with blood from the chest down.

To finish it, to make absolutely sure, he found a rock and pounded what was left of Cripps' leg through the Guernsey's right ear. The animal bucked weakly, flopping against the ground, then farted in one long gust and dribbled shit between its splayed thighs.

Steven spread his arms and stretched. He could reclaim the sleeping herd now, he had the strength to reclaim a thousand herds. Under the arc-light of an ultimate selfishness the path before him led straight to a future where there would always be safety and something to be loved by – where he would always have a family.

Nothing he did not allow could enter here, and in this world, free of comparison with the outside, he would live a life as perfect as any he had seen on TV.

The roan nuzzled his side, he dropped his arm around her. Already he could feel the wind against his skin, the unstoppable power of the stampede, see the tracering splash of dim lights and the rush of tunnel walls, feel the glory of motion and power, his expansion into being... and the blessed communion of belonging.

He filled his lungs to shout. It was time to wake the herd.